EXTREME WEATHER

Edited by **Philip John Sallis**

Extreme Weather
http://dx.doi.org/10.5772/intechopen.72136
Edited by Philip John Sallis

Contributors

Ali Assani, Chi-Cherng Hong, Yi-Kai Wu, Carlos Navia, Marcel De Oliveira, Carlos Augusto, Yuqing Wang, Zifeng Yu, Andriy Mykhnovych, Vira Balabukh, Olha Pylypovych, Olena Lavrynenko, Volodymyr Bilaniuk, Polioptro F. Martinez-Austria, Erick R. Bandala, Ephias Makaudze

First published in London, United Kingdom, 2018 by IntechOpen
IntechOpen is the global imprint of INTECHOPEN LIMITED, registered in England and Wales, registration number: 11086078, The Shard, 25th floor, 32 London Bridge Street
London, SE19SG – United Kingdom
Printed in Croatia

British Library Cataloguing-in-Publication Data
A catalogue record for this book is available from the British Library

Additional hard copies can be obtained from orders@intechopen.com

Extreme Weather, Edited by Philip John Sallis
p. cm.
Print ISBN 978-1-78923-612-5
Online ISBN 978-1-78923-613-2

For EU product safety concerns:
IN TECH d.o.o., Prolaz Marije Krucifikse Kozulić 3, 51000 Rijeka, Croatia,
info@intechopen.com or visit our website at intechopen.com.

Meet the editor

Philip's career spans some 45 years of academic positions in the UK, Australia, and New Zealand, with adjunct positions in Hong Kong, the USA, and Chile. After his NZ undergraduate studies in Mathematics and Computer Science, he completed his PhD degree from the University of London (City) in 1979. Returning to NZ in 1987, he was appointed to the foundation chair in Information Science at the University of Otago. In 1999, he was appointed as the deputy vice chancellor at the Auckland University of Technology (AUT), and since 2009, he has led a research group there in the field of GeoComputation with wireless sensor network and agrometeorological projects in nine countries, particularly Chile, Germany, and Japan. During this time, he also held senior university management roles as a pro-vice chancellor. From 2018, he has reduced his full-time position to a part-time role as a research professor at the School of Engineering, Computer, and Mathematical Sciences, AUT.
He is a fellow of the Institute of Information Technology Professionals in New Zealand and a member of the Royal Society.
He is a life member of the International Association of Mathematical Geophysics and a senior member of the IEEE. He recently received Life Membership Award of that institution.

Contents

Preface

The term *extreme weather* normally conjures up thoughts of massive storms or heat waves or overtly cold temperatures. These are all examples of what we might consider as weather events that occur out of the ordinary or what is regarded as the normal pattern of calm, heat, cold, dry, or wet conditions for one season of the year or another. The point is that if we consider an oscillation of data points in a weather pattern and plot a mean through it, extreme weather can be observed as a perturbation in a distribution of climatic events over time. These events may be short-lived, such as a wind gust occurrence, or of longer duration, such as heavy rain leading to flooding. Importantly, once initiated, a perturbation event has an associated consequence, which usually requires human intervention to rectify the event's consequences.

Each of the chapters in this book identifies and describes an extreme weather event or collection of events and their consequences. Mostly, these events are related to a specific location, but general points are drawn out from the specific causes and effects discussed.

The chapter on *Heat Waves* is a good example of this where trends are observed across a set of example events, with their effects and influence on climatic change described. In this detailed discussion of heat waves as extreme weather events, dangerous temperature levels are considered with their effects on vegetation and water.

As the result of northwest winds over the Korean peninsula, extratropical cyclones occur in the East Sea of Japan. The chapter here on this topic provides a fascinating insight into what are discrete extreme weather events, but even so, they are expected as a cause-and-effect scenario and, therefore, can be predicted. The effect or the consequence of the huge waves that are formed from these cyclones can be devastating, so forecasting their formation and arrival when they make land is vital to the people living in the coastal areas of Eastern Japan for their personal safety and the survival of their farms and businesses.

Some aspects of the cyclone phenomenon are described in the chapter on the influence of abrupt changes in weather to climate regimes in specific locations. The chapter argues that thermal conditions play a more dominant role in influencing climate shifts than dynamic events over time. It also takes a wider view of the longer-term prevailing climate for regions where cyclones are experienced.

A more calm and yet potentially just as devastating weather event is regularly experienced in the mid-west of the United States. Again, these events are expected, so it can to some extent be predicted. In the case study chapter describing ice cover on the Grand River in Ohio, we read about a slower-moving weather event but with the consequence of severe flooding that can occur at any time after the ice begins to melt. This chapter discusses the environmental impact of this extreme weather example and reinforces the point that not all such events occur suddenly or without warning.

On the other hand, more dynamic extreme weather events and their consequences, such as those described in the chapter on *Geomagnetic Storms*, are difficult to predict due to the short buildup time duration, similar to wind gusts. Very often these weather events have a destructive consequence. The solar wind velocities can reach up to 500 km per second. Instruments for measuring the events and their variables are described in the chapter with a discussion of issues relating to data collection and analysis.

Together, these chapters provide a sound basis for considering the intrinsic elements of extreme weather events, and the place of the topics addressed within the chapters can assist in the accumulated knowledge being built to better understand the weather and climate dynamics.

Professor Philip Sallis
School of Engineering, Computer and Mathematical Sciences
Auckland University of Technology
New Zealand

Chapter 1

Comparison of the Temporal Variability of Maximum Daily Temperatures for Summer Months in Relation to El Nino Events in Southern Québec

Ali Assani

Additional information is available at the end of the chapter

http://dx.doi.org/10.5772/intechopen.74548

Abstract

The goals of this study were (1) to compare the long-term trend of the interannual variability of maximum daily temperatures for four summer months (June, July, August, and September) using the Spearman's rank correlation coefficient and Mann-Kendall tests and (2) to analyze the link between these temperatures and El Niño events of varying intensities using the linear correlation method. Data from 23 stations for the period from 1950 to 2010 were analyzed. As far as the analysis of the long-term trend is concerned, the observed warming is greater for the last 2 months (August and September) than for the first 2 months (June and July) of the summer season, likely as a result of the warming of ocean surface waters. As for the link between El Niño events and summer maximum daily temperatures, a negative correlation was highlighted for the first time between these two variables for southern Quebec. However, this correlation is only observed for the two "cooler" summer months (June and September), likely due to a weak influence of site (station) characteristics on maximum daily temperature variations.

Keywords: maximum daily temperatures, summer, El Niño events, long-term trend, correlation, southern Quebec

1. Introduction

It is a well-established fact that, in cool temperate regions, the extent of climate warming will vary seasonally. Several studies have compared the interannual variability of temperatures between seasons in Canada (e.g., [1–7]). These studies have shown that, since 1950, climate warming is more pronounced in winter and spring than in summer and fall. However, Guerfi et al. [8]

showed that, in Quebec, climate warming is not as strong in winter, with very few stations analyzed recording a significant increase in maximum and minimum daily temperatures since 1950. Assani et al. [9] showed that, in Quebec, climate warming is more generalized and greater in summer than in winter. Thus, unlike what is observed for winter, the significant increase in summer maximum and minimum daily temperatures since 1950 is observed at many stations. Moreover, the observed nighttime warming in summer is greater than the observed daytime warming. These studies, however, only looked at variability between seasons, and do not address interannual variability for individual seasons (intraseasonal variability) in order to see if the extent of warming is the same or not for each month of a given season. The first objective of this study is, therefore, to compare the long-term trend of the interannual variability of maximum daily temperatures for the four months (June, July, August, and September) that make up the summer season in Quebec. The underlying hypothesis is that the increase in temperature is similar for the four summer months in southern Quebec.

In addition, analysis of the relationship between large-scale climate oscillations and summer climate in Canada has revealed that the links between these variables are relatively weak and variable compared to what is observed for the same variables in winter (e.g., [10]). However, no study has looked at the relationship between El Niño events and summer temperatures in Quebec. El Niño events affect temperatures at the global scale, and intense El Niño events are generally associated with relatively high temperatures (positive anomalies) at the global scale (e.g., [11]). The question is, therefore, whether there is a link between the intensity of El Niño events and maximum daily temperatures during the four summer months in Quebec, something that has not been addressed before.

2. Methods

2.1. Choice of stations

Twenty-three stations were selected to analyze the temporal variability of summer (June–September) temperature and rainfall (**Table 1** and **Figure 1**). The selection of these stations was primarily based on the availability of continuous measurements of temperature over the period from 1950 to 2010. This period was selected in order to analyze the largest possible number of stations. Prior to 1950, there were very few stations in Quebec where temperature and rainfall was measured. Since 2000, to streamline the measurement of hydroclimate variables in Canada and Quebec, many weather and hydrometric stations were shut down. For this study, all stations for which hydroclimate variables have been measured quasi-continuously for at least 50 years during the period from 1950 to 2010 were used. Summer temperature and rainfall data were taken from the Environment Canada website (http://www.climate.weatheroffice.ec.gc.ca/, viewed in March 2016). For each summer month, a series of maximum daily temperatures was produced.

As far as El Niño events are concerned, only events of differing intensities (weak, moderate, strong, and very strong) following the NOAA classification ([12]) were selected. From 1950 to 2010, 22 El Niño events were recorded, including 11 events of weak intensity

http://dx.doi.org/10.5772/intechopen.74548

Stations	Code	Years	ID	Latitude (N)	Longitude (W)	Altitude (m)
Coaticook	01	1950–2010	7021840	45°09′	71°48′	259
Les Cèdres	02	1950–2010	701429	45°18′	74°03′	47
Magog	03	1950–2010	7024440	45°16′	72°07′	274
Montréal (Trudeau)	04	1950–2010	7025250	45°28′	73°45′	36
Oka	05	1951–2010	7015730	45°30′	74°04′	91
Philipsburg	06	1950–2010	7026040	45°02′	73°05′	53
St Ephrem	07	1950–2010	7027200	46°04′	70°58′	312
St-Malo d'Auckland	08	1950–2010	7027520	45°12′	71°30′	564
Ste Anne de la Pérade	09	1950–2010	7016840	46°38′	72°14′	16
Valleyfield	10	1950–2010	7028680	45°17′	74°06′	46
Bagotville	11	1950–2010	7060400	48°20′	71°00′	159
Mont Joli A	12	1950–2006	7055120	48°36′	68°13′	48
Natashquan A	13	1950–2002	7045410	50°11′	61°48′	575
Sept îles A	14	1950–2001	7047910	50°13′	66°16′	310
Ste Rose de Dégelis	15	1950–2001	7057720	47°34′	68°38′	151
Trois-Pistoles	16	1950–2001	7058560	48°09′	69°07′	58
Chelsea	17	1950–2010	7031660	45°31′	75°47′	112
La Tuque	18	1950–2001	7074240	47°24′	72°47′	152
Nicolet	19	1950–2010	7025440	46°12′	72°37′	30
Nominingue	20	1950–2010	7035520	46°23′	75°03′	305
St Alban	21	1950–2010	7016800	46°43′	72°05′	76
St Jérôme	22	1950–2010	7037400	45°48′	74°03′	169
Milan	23	1950–2010	7024920	45°37′	71°07′	482

Table 1. Some data from analyzed stations.

(1951–1952, 1952–1953, 1953–1954, 1958–1959, 1968–1969, 1969–1970, 1976–1977, 1979–1980, 1994–1995, 2004–2005, and 2006–2007), 6 events of moderate intensity (1963–1964, 1986–1987, 1987–1988, 1991–1992, 2002–2003, and 2009–2010), 3 events of strong intensity (1957–1958, 1965–1966, and 1972–1973), and 2 events of very strong intensity (1982–1983 and 1997–1998). Based on Oceanic Niño Index (ONI, the running 3-month mean SST anomaly for the Niño 3.4 region) values, an event is considered to be an El Niño event when the ONI values for five consecutive overlapping 3-month periods are at or above the +0.5° anomaly. The threshold is further broken down into weak (with a 0.5–0.9 SST anomaly), moderate (1.0–1.4), strong (1.5–1.9), and very strong (≥2.0) events [12]. However, for this study, rather than using ONI values, Southern Oscillation Index (SOI) values were used because they are available and widely used in the scientific literature. Moreover, using SOI values makes it

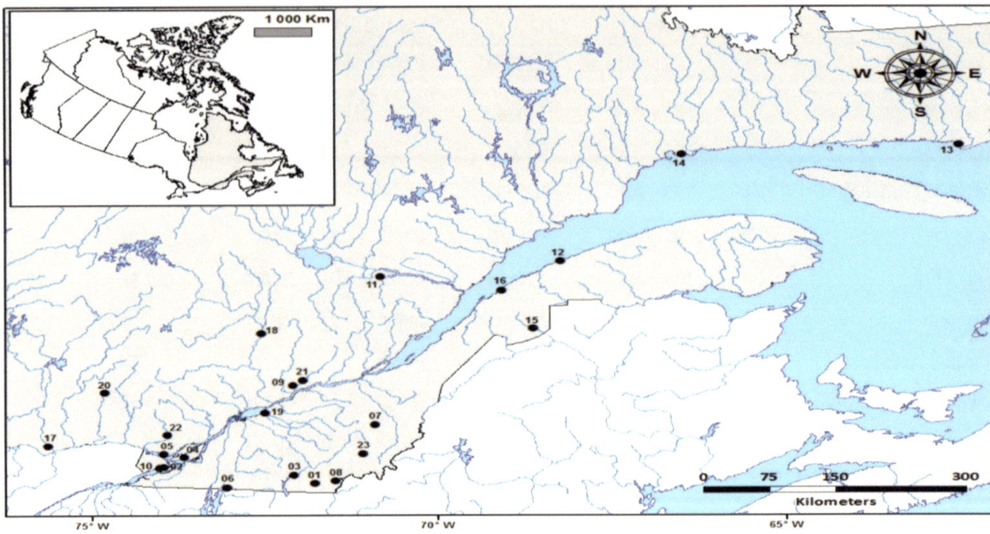

Figure 1. Location of stations analyzed.

possible to readily compare the results of this study with those of previous work. It is, however, important to note that SOI values associated with the different El Niño intensities are negative because they are derived from differences in pressure, whereas ONI values are measurements of surface water temperatures of the tropical Pacific Ocean. Be that as it may, the two types of indices (SOI and ONI) lead to the same results due to their strong correlation.

For each of the 22 El Niño events, the following six seasonal mean values of SOI indices were then derived: AMJ-1 is the mean SOI value for the months of April, May, and June of the previous years; JAS-1 is the mean SOI value for the months of July, August, and September of the previous year; OND-1 is the mean SOI value for the months of October, November, and December of the previous year; JFM is the mean SOI value for the months of January, February, and March of the current year; AMJ is the mean SOI value for the months of April, May, and June of the current year; and JAS is the mean SOI value for the months of July, August, and September of the current year. SOI index data were taken from the NOAA website (National Oceanic and Atmospheric Administration, http://www.cdc.noaa.gov/ClimateIndices/List, viewed on November 23, 2017). Indices for the previous year, when the El Niño event is ongoing in the Pacific Ocean, were included because the influence of El Niño events on climate in Quebec is observed after a lag of at least 6 months.

2.2. Statistical analysis

Statistical analysis of data was conducted in two steps. In the first step, the Spearman's rank correlation coefficient and Mann-Kendall tests were used to constrain the long-term trend of the temporal variability of maximum daily temperatures for the four summer months. Both tests are widely used in climate science and have been described in many articles (e.g., [13]). Because both tests yielded the same results, we only present results obtained using the Spearman's rank coefficient test. It is important to note that the series analyzed did not show any autocorrelation.

Spearman's correlation coefficient is calculated using the following equation [13]:

$$r_s = \frac{\sum i\, y_i - (\sum i)^2/n}{s_i\, s_{yi}} \tag{1}$$

$$s_i^2 = \sum i^2 - (\sum i)^2/n \quad \text{and } s_{yi}^2 = \sum y_i^2 - (\sum i)^2/n \tag{2}$$

where n is the sample size (number of years during which streamflow was measured) and *yi* is the rank of the value of streamflow measured in year i for streamflow values ranked in order of increasing magnitude.

The test critical value is calculated using the following equation:

$$t_s = r_s \sqrt{\frac{n-2}{1-r_s^2}} \tag{3}$$

The null hypothesis (absence of a significant trend) is rejected if the value of t_s is higher than the value appearing in the table of Student's t values for a given probability level with n-2 degrees of freedom.

The last step consisted of correlating maximum daily temperatures for the four summer months with the six SOI series for the 22 El Niño events of differing intensities. It is important to note that years that do not match the years of occurrence of an El Niño event were excluded from this analysis so as to only look at the influence of El Niño events on the interannual variability of summer maximum daily temperatures. No previous study has considered El Niño events exclusively. In this study, only the simple correlation method was used because multivariate analysis methods (canonical correlation analysis and redundancy analysis) did not yield conclusive results due to the relatively small sample size (22).

3. Results

3.1. Comparison of the long-term trend of maximum daily temperatures for the four summer months

Results obtained using the Spearman's rank correlation coefficient test is presented in **Table 2**. Very little change in the long-term trend is observed for June and July, with only three stations recording changes for these two months. In contrast, changes in the long-term trend are observed at numerous stations (more than 40% of stations analyzed) for August and September. For all four summer months, these changes reflect a positive long-term trend or a significant increase in maximum temperatures over time. Thus, statistically significant values of rare all positive (see **Figure 2**). These results are identical to those obtained using the Mann-Kendall test.

Stations	June		July		August		September	
	r_s	t_s	r_s	t_s	r_s	t_s	r_s	t_s
Coaticook	−0.1386	1.0750	−0.1887	1.4760	−0.0015	0.0112	0.1217	0.9419
Les Cèdres	0.1416	1.0988	0.1899	1.4855	**0.2854**	**2.2870****	**0.2548**	**2.0242****
Magog	−0.0615	0.4739	−0.0512	0.3940	0.1266	0.9804	**0.2542**	**2.0191****
Montréal (Trudeau)	0.0580	0.4462	0.0902	0.6958	0.1976	1.5481	**0.2602**	**2.0696****
Oka	0.0841	0.6428	0.0307	0.2339	0.1534	1.1820	**0.2419**	**1.8989***
Philipsburg	−0.0232	0.1782	−0.0173	0.1330	0.1329	1.0298	0.1817	1.4195
St Ephrem	0.1794	1.4000	0.1734	1.3525	**0.4047**	**3.3993*****	**0.3172**	**2.5695****
St-Malo d'Auckland	**0.2887**	**2.3165****	0.0883	0.6811	**0.3743**	**3.1000*****	0.2115	1.6620
Ste Anne de la Pérade	−0.0017	0.0128	−0.0992	0.7662	0.1803	1.4077	0.2206	**1.7376***
Valleyfield	−0.0861	0.6581	−0.0143	0.1087	0.0701	0.5357	0.1032	0.7902
Bagotville	0.1750	1.3656	0.1961	1.5360	**0.3212**	**2.6055****	**0.3231**	**2.6222****
Mont Joli A	**0.2428**	**1.8562***	0.1761	1.3267	**0.3751**	**3.0010*****	**0.3231**	**2.5317****
Natashquan A	**0.2643**	**1.9572****	**0.3389**	**2.5724****	**0.5119**	**4.2550*****	**0.5245**	**4.4000*****
Sept Îles A	0.0601	0.4256	−0.0199	0.1405	0.0882	0.6264	−0.0404	0.2859
Ste Rose de Dégelis	0.2388	1.7390	0.0272	0.1930	**0.2661**	**1.9518****	−0.1369	0.9771
Trois-Pistoles	0.1110	0.7896	0.0117	0.0830	**0.2447**	**1.7849***	−0.1021	0.7264
Chelsea	0.1468	1.1398	0.1838	1.4359	**0.2940**	**2.3626****	**0.2725**	**2.1758****
La Tuque	0.1318	0.9405	−0.1172	0.8348	0.0913	0.6482	−0.1772	1.2733
Nicolet	0.1210	0.9361	−0.0954	0.7364	0.0672	0.5174	0.1750	1.3654
Nominingue	−0.0862	0.6648	0.0023	0.0174	0.1605	1.2491	0.1373	1.0645
St Alban	0.1903	1.4890	0.1995	1.5640	**0.4039**	**3.3913*****	**0.3668**	**3.0282*****
St Jérôme	0.0142	0.1089	0.0084	0.0646	0.1933	1.5131	0.1200	0.9287
Milan	0.0436	0.3349	0.1632	1.2706	**0.2796**	**2.2369****	0.1953	1.5294

*Statistically significant value at the 10% level.

**Statistically significant value at the 5% level.

***Statistically at the 1% level.

The statistically significant values are shown in the bold.

Table 2. Analysis of the long-term trend of monthly maximum daily temperatures in summer using the Spearman's rank correlation coefficient method.

3.2. Relationship between El Niño events and maximum daily temperatures for the four summer months

Coefficient of correlation values are presented in **Table 3**. Only the AMJ-1 and JAS-1 SOI values associated with El Niño events are significantly correlated with maximum daily temperatures.

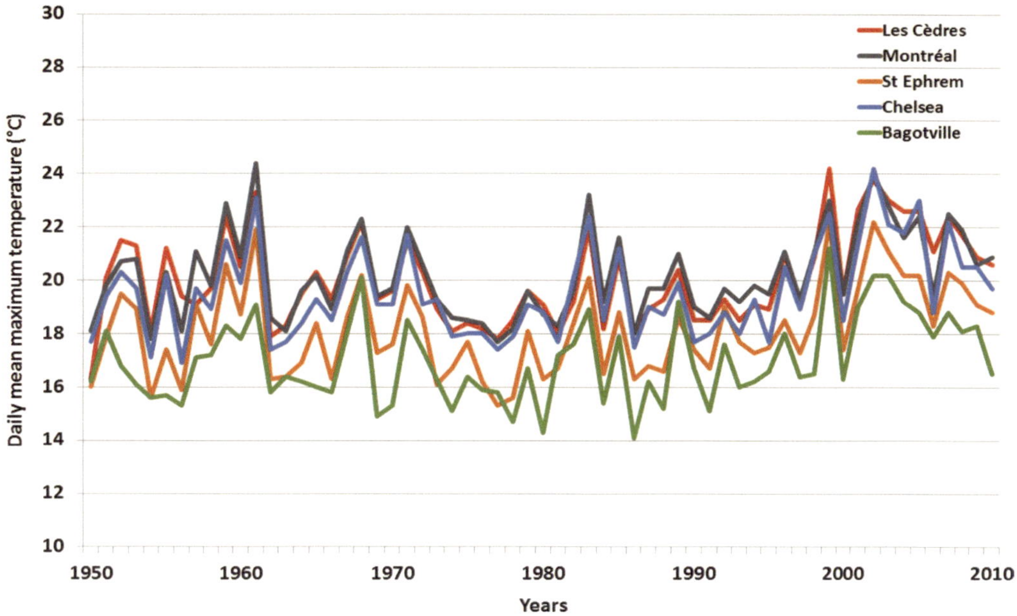

Figure 2. Example of the interannual variability of maximum daily temperatures in August at the five stations.

Stations	June		July		August		September	
	AMJ-1	JAS-1	AMJ-1	JAS-1	AMJ-1	JAS-1	AMJ-1	JAS-1
Coaticook	−0.1578	**−0.4229***	0.1142	−0.0181	0.0393	−0.0484	0.1897	−0.2378
Les Cèdres	−0.1889	−0.1544	−0.0280	0.0392	−0.0391	−0.0786	0.1130	−0.2273
Magog	−0.1486	−0.2777	0.0966	0.0594	0.0233	−0.0141	0.2079	−0.1615
Montréal (Trudeau)	−0.2117	−0.2669	−0.0734	−0.0871	−0.0002	−0.0588	0.0310	−0.3406
Oka	−0.2892	−0.3465	−0.1437	−0.0270	−0.0474	−0.0781	−0.0137	**−0.3906***
Philipsburg	−0.2595	−0.2013	0.0904	0.0198	0.1249	0.1328	−0.0474	**−0.3620***
St Ephrem	**−0.3941***	**−0.4142***	−0.1213	−0.1357	−0.0621	−0.1164	0.1026	−0.2777
St-Malo d'Auckland	−0.2322	−0.3153	−0.0176	−0.0052	−0.0089	−0.0296	0.1948	−0.1651
Ste Anne de la Pérade	**−0.4070***	**−0.4218***	−0.0825	−0.0085	−0.0383	−0.1160	−0.0207	**−0.3662***
Valleyfield	−0.2639	−0.2861	0.1386	−0.0373	0.0839	0.0769	0.1323	0.2561
Bagotville	−0.3206	−0.3052	−0.1385	−0.0618	−0.2458	**−0.3806***	−0.1456	−0.3262
Mont Joli A	−0.2874	**−0.4049***	−0.1402	0.0169	−0.3218	−0.2273	−0.0543	**−0.4301***
Natashquan A	0.0502	−0.0133	−0.3629	−0.1946	−0.2454	0.0236	−0.4491	**−0.6531***
Sept Îles A	−0.3191	−0.2441	−0.0870	0.1571	−0.1390	−0.0413	0.0187	**−0.3861***
Ste Rose de Dégelis	**−0.3791***	**−0.5610***	−0.1091	−0.0026	−0.1689	−0.1560	0.0176	**−0.4217***

Stations	June		July		August		September	
	AMJ-1	JAS-1	AMJ-1	JAS-1	AMJ-1	JAS-1	AMJ-1	JAS-1
Trois-Pistoles	**−0.3804***	**−0.3627***	−0.1865	−0.0088	−0.2863	−0.1797	−0.1605	**−0.4842***
La Tuque	**−0.4994***	−0.2942	−0.1440	−0.0798	−0.1968	−0.3281	−0.1425	**−0.4572***
Nicolet	**−0.4994***	−0.2942	−0.1440	−0.0798	−0.1968	−0.3281	−0.1425	**−0.4572***
Nominingue	**−0.5110***	**−0.5678***	0.0272	0.0702	0.1166	0.2012	0.1466	−0.0546
St Alban	−0.2803	**−0.3733***	0.0424	−0.0856	0.2354	−0.0953	0.1877	−0.1157
St Jérôme	**−0.3717***	−0.3599	−0.2435	−0.3010	−0.0073	−0.1750	0.0529	−0.3200
Milan	**−0.4178***	**−0.4380***	−0.1107	−0.1471	−0.1157	−0.1191	0.1206	−0.2606

*Statistically significant value at the 5% level.
The statistically significant values are shown in the bold.

Table 3. Coefficients of correlation calculated between El Niño events and monthly maximum daily temperatures in summer in Quebec (1950–2010). .

There is no significant correlation for the other four SOI values. No statistically significant correlation was found between El Niño events and maximum daily temperatures in July, and such a correlation is only observed for one station for the month of August. In contrast, a correlation is found at nearly half of the stations for the months of June and September, the two coolest summer months. All statistically significant correlations are negative. Thus, maximum daily temperatures tend to increase with the intensity of El Niño events. It is worth pointing out that such a relationship is not observed with La Niña events (results not presented herein). Also of note, temperatures in June are correlated with the AMJ-1 and JAS-1 climate indices, whereas temperatures in September are only correlated with the JAS-1 index (**Table 3**).

4. Discussion and conclusion

Comparison of the long-term trend of maximum daily temperatures for four summer months (June, July, August, and September) using the Spearman and Mann-Kendall tests highlighted a statistically significant increase in temperature in southern Quebec over the period from 1950 to 2010. However, this increase is not uniform for the four months. The warming trend is observed at nearly half of the stations analyzed for the months of August and September, but at fewer than 15% of stations for the months of June and July. The first main conclusion arising from the study is the fact that the last two months of the summer season are warming faster than the first two months. For July, a warming trend is only observed at one station. This suggests that the warm summer season tends to end late in southern Quebec. It was not possible to constrain the potential climate factors that could account for the warming trend for the months of September and August. One hypothesis is the increasingly widely observed warming of ocean surface waters, in the Atlantic and Pacific Oceans among other places, the influence of which on temperature in continental North America has been highlighted by a number of authors (e.g., [11, 14]). It should be recalled that, due to thermal inertia, the warming of ocean surface waters accelerates in August in the northern hemisphere.

There is a widespread assumption that, in Canada in general and in Quebec in particular, El Niño events have no impact on summer climate. This assumption is not always based on an analysis of the relationship between climate variables and these events. The analysis carried out in this study revealed the existence of a significant negative correlation between maximum daily temperatures and El Niño events of varying intensities during the period from 1950 to 2010. Like the long-term trend of the interannual variability of temperature, this correlation varies between the four summer months. It is not observed for the months of July and, to a lesser extent, August, but is observed at nearly half of the stations analyzed for June (early summer) and September (late summer), the two coolest summer months in southern Quebec. The lack of correlation between El Niño events and maximum daily temperatures in July and August, the two warmest months of summer, may be explained by the fact that, as a result of relatively substantial warming during these two months, the influence of local site (station) characteristics on daily daytime temperature variations becomes more important (microclimate effect) than that of regional climate factors (air mass circulation and fronts). Thus, during the warmest part of summer, local warming enhances the strong spatial variability of convective movements that affect daytime temperatures in general and maximum temperatures in particular. These convective movements produce convective clouds and localized and dispersed storms, thus inducing differences in maximum temperatures between sites.

The negative correlation highlighted between maximum daily temperatures and El Niño events suggests that an increase in the intensity of these events in the current climate warming context, as predicted by many climate models, would lead to an increase in maximum daily temperatures particularly in June and September in southern Quebec. Thus, the four summer months will become increasingly warmer in southern Quebec.

Author details

Ali Assani

Address all correspondence to: ali.assani@uqtr.ca

Department of Environmental Sciences, University of Québec at Trois-Rivières, Québec, Canada

References

[1] Casati B, de Elia R. Temperature extremes from Canadian Regional Climate Model (CRCM) climate change projections. Atmosphere-Ocean. 2014;**52**(3):191-210

[2] Mekis E, Vincent LA, Shepard MW, Zhang X. Observed trends in severe weather conditions based on humidex, wind chill, and heavy rainfall events in Canada for 1953-2012. Atmosphere-Ocean. 2015;**53**(4):383-397

[3] Turner JK, Gyakum JR. Trends in Canadian surface temperature variability in the context of climate change. Atmosphere-Ocean. 2010;**48**(3):147-162

[4] Vincent LA, van Wijngaarden WA, Hopkinson R. Surface temperature and humidity trends in Canada for 1953-2005. Journal of Climate. 2007;**20**:5100-5113

[5] Wang XL, Feng Y, Vincent LA. Observed changes in one-in-20 year extremes of Canadian surface air temperatures. Atmosphere-Ocean. 2014;**52**(3):222-231

[6] Yagouti A, Boulet G, Vincent L, Vescocvi L, Mekis E. Observed changes in daily temperature and precipitation indices for southern Québec, 1960-2005. Atmosphere-Ocean. 2000;**3**:395-429, 46(2);243-256

[7] Zhang X, Vincent LA, Hogg WD, Niitsoo A. Temperature and precipitation trends in Canada during the 20th century. Atmosphere-Ocean. 2000;**3**:395-429

[8] Guerfi N, Assani AA, Mesfioui M, Kinnard C. Comparison of the temporal variability of winter daily extreme temperatures and precipitations in southern Quebec (Canada) using the Lombard and copula methods. International Journal of Climatology. 2015; **35**:4237-4246

[9] Assani AA, Maloney-Dumont V, Pothier-Champagne A, Kinnard C, Quéssy J-F. Comparison of the temporal variability of summer temperature and rainfall as it relates to climate indices in southern Quebec (Canada). Theoretical and Applied Climatology. 2017 (under review)

[10] Bonsal B, Shabbar A. Oscillations climatiques à grande échelle ayant une incidence sur le Canada, de 1900 à 2008. Biodiversité canadienne : état et tendances des écosystèmes en 2010. Rapport technique thématique n°4. Ottawa (ont.): Conseils canadiens des ministres des ressources; 2011. iii+15 p http://www.biodivcanada.ca/default.asp?lang=Fr&n=137E1147-1

[11] Trenberth KE, Fasullo JT. An apparent hiatus in global warming. Earth's Future. 2013; **1**:19-32

[12] Null J. El Niño and La Niña years and intensities based on Oceanic Niño index (ONI). 2016. http://ggweather.com/enso/oni.htm. [Viewed December 14, 2016]

[13] Sneyers R. Sur l'analyse statistique des séries d'observations. Note technique n°143. Genève: Organisation Météorologique Mondiale; 1975. 192 p

[14] Sutton RT, Hodson DLR. Atlantic Ocean forcing of North American and European summer. Climate Sciences. 2005;**309**:115-118

Chapter 2

Influence of Climate Regime Shift on the Abrupt Change of Tropical Cyclone Activity in Various Genesis Regions

Chi-Cherng Hong and Yi-Kai Wu

Additional information is available at the end of the chapter

http://dx.doi.org/10.5772/intechopen.74947

Abstract

In this chapter, we reported the effect of basin-scale climate regime shift (CRS) on the abrupt change of tropical cyclone (TC) activity in various genesis basins, including the Pacific, Atlantic, and Indian Oceans. An analysis of regime shift index reveals that the worldwide TC activity experienced four significant abrupt changes during 1960–2014, including (i) an abrupt increase/decrease in the eastern North Pacific (ENP)/western North Pacific (WNP) in the early 1970s, (ii) an abrupt increase in the ENP and WNP in the early 1980s, (iii) an abrupt increase in the North Atlantic and ENP in the middle 1990s, and (iv) an abrupt decrease in the WNP and western South Pacific in the late 1990s. Three of them are identified concurrent with a significant CRS. The possible influence of a CRS on the abrupt change of TC activity in various genesis regions is addressed. We demonstrate that a CRS induced time mean state shift results in a rapid change in the large-scale dynamic and thermodynamic conditions, which substantially contributes to the abrupt change of TC activity in various genesis regions. In addition the CRS, the effect of interdecadal variability, such as the interdecadal Pacific Oscillation and Atlantic Multidecadal Oscillation, on the abrupt change of TC activity was discussed.

Keywords: climate regime shift, abrupt change, tropical cyclone, interdecadal Pacific Oscillation, Atlantic Multidecadal Oscillation

1. Introduction

The activity of tropical cyclones (TCs) substantially affected by large scale thermal and dynamic conditions fluctuates with multiple time scales. For example, the TC genesis number in the western North Pacific (WNP), the TC occurs most frequently in this region of the worldwide, exhibits an interannual and interdecadal fluctuation [1–6]. The interannual variation of TC activity in the WNP has been shown to be associated with large scale circulation anomalies caused by El Niño and Southern Oscillation [2, 7], as well as by stratospheric quasi-biennial oscillation [6, 8]. However, thermal conditions play a more dominant role than dynamic conditions in the interannual variation of TC activity in the North Atlantic [9–11].

In the view of interdecadal time scale, the TC genesis frequency in the WNP was high in mid-1960s and late 1980s but relatively low in 1970s and late 1990s [8, 12, 13]. It was noted a significant decrease in TC activity over the WNP after 1998 [8]. They attributed this decrease to the change in large scale vertical wind shear and 500-hPa geopotential height fields, which is likely associated with Pacific decadal Oscillation (PDO). In contrast to WNP, the TC activity in the North Atlantic Ocean showed an interdecadal increase since mid-1990s [10, 14, 15], which is likely a response to the interdecadal warming in the tropical Atlantic due to the Atlantic Multidecadal Oscillation (AMO) [10, 15].

Whereas the interdecadal variation of TC activity in various genesis regions had been widely discussed, most of previous studies are focused on specific genesis region. The TC activity in various genesis regions possibly had experienced an interdecadal change simultaneously, and so far, this issue was not thoroughly investigated yet. Recently, Hong et al. [16] reported that the tropical Pacific Ocean experienced a basin-scale climate regime shift (CRS) during the middle to late 1990s (approximately 1996/1997). This CRS, characterized by a negative sea surface temperature anomaly (SSTA) in the tropical central Pacific, generated a pair of anticyclone in the tropical western Pacific and that resulted in the interdecadal decrease of TC activity in the WNP and western South Pacific (WSP) in the late 1990s [17]. In this chapter, we extended the previous work [17], the possible influence of a CRS on the abrupt change of TC activity in various regions was investigated. Because a CRS is generally accompanied with atmospheric or oceanic low frequency variability, such as PDO and interdecadal Pacific Oscillation (IPO), the contribution of low frequency variability of climate system on the abrupt change of TC activity was discussed.

2. Data and methodology

Data from the International Best Track Archive for Climate Stewardship (IBTrACS; [18]) from the Regional Specialized Meteorological Center of the World Meteorological Organization and various agencies (e.g., Joint Typhoon Warning Center, JTWC, and the Japan Meteorology Agency, JMA) during the period of 1960–2014 were used to analyze TC frequency change. Because the observation of tropical depression-associated wind speed was not available until 1980 in the southern hemisphere, the analysis of TC activity in the Indian Ocean (IO) and western South Pacific (WSP) starts from the 1980s. Besides, the typhoon season (October to April) in the southern hemisphere covers two calendar years, the TC genesis number was measured from July to next June. Monthly atmospheric datasets of NCEP/NCAR Reanalysis I [19] and monthly SST field data obtained from the Met Office Hadley Centre [20] were used for diagnosing large scale circulation anomalies. The TC is defined as the maximum sustained wind speed of tropical cyclone reaches 34 knots. The TC is further separated to three scales, TS,

http://dx.doi.org/10.5772/intechopen.74947

TY and ST, which is approximately corresponding to the TS, C1–C2, C3–C5 of Saffir-Simpson typhoon-scale [21] respectively (**Table 1**). A singular vector decomposition (SVD) analysis by pairing up the SST and 850 hPa stream function fields was conducted to investigate the climate regime shift (CRS) of coupled climate system. Because the global SST, especially in the Indian Ocean (IO), exhibits a remarkable warming trend, the linear trend of SST during 1950–2010 had been removed off in the SVD analysis. A regime shift index (RSI; [22]) was them applied to the time series of first two SVDs to detect the specific time of climate regime shift. A cut-off length of 10 years was applied for climate regime shift (CRS) diagnosis, and only the RSI exceeding significant test at 0.05 significance level was considered.

3. Abrupt change of TC activity in various genesis regions

Figure 1 shows the time series of the annual TC count in various genesis basins. The TC genesis number in the WNP shows an interdecadal variation, indicating that the TCs were active approximately in the periods of 1960–1970 and 1980–1997, but were inactive in 1970–1980 and 1998–2014. This interdecadal variation of TC activity was consistent with the previous study [8]. The RSI calculation shows that the TC number underwent three significant abrupt changes in 1972, 1983 and 1997 during the 55-year period (1960–2014). Note that the RSI value in 1983 is in the margin of 5% significant level, and therefore is not clearly seen in **Figure 1**. Among the RSI, an abrupt decrease in TC activity after 1998 was particularly pronounced; the annual mean TC number was near 30 in the active period of 1988–1997, but decreased dramatically to 25 in the inactive period of 1999–2008. Time series of different TC scales (**Figure 2a**) reveals that the abrupt drop of TC number after 1998 is primarily caused by the decrease in TY. The number of TS and ST do not exhibit significant change prior and after the change point in 1998.

The TC genesis number in the eastern North Pacific (ENP) was low in 1960–1970 but was high in 1980–1990 and was approximately normal in 1970–1980 and after mid-1990s (**Figure 1**). The RSI calculation shows that the TC number in the ENP underwent three significant abrupt changes in 1969, 1981, and 1992 during 1960–2014. **Figure 2** further shows that the number of

Type of TC	Sustained wind speed (units: Knots)	Saffir-Simpson hurricane wind scale
Td	17–33	Tropical depression
TS	34–63	Tropical storm
TY	64–104	C1 (64–82) C2 (83–95)
ST	≧105	C3 (96–112) C4 (113–136 t) C5 (≧137)

The TY (typhoon) is approximately corresponding to category 1 + 2, and ST (sever typhoon) is corresponding to category 3 + 4 + 5 of Saffir-Simpson hurricane wind scale [21]. The TC number is defined as the summation of TS, TY, and ST numbers.

Table 1. Definition of the different tropical cyclone scale in this study.

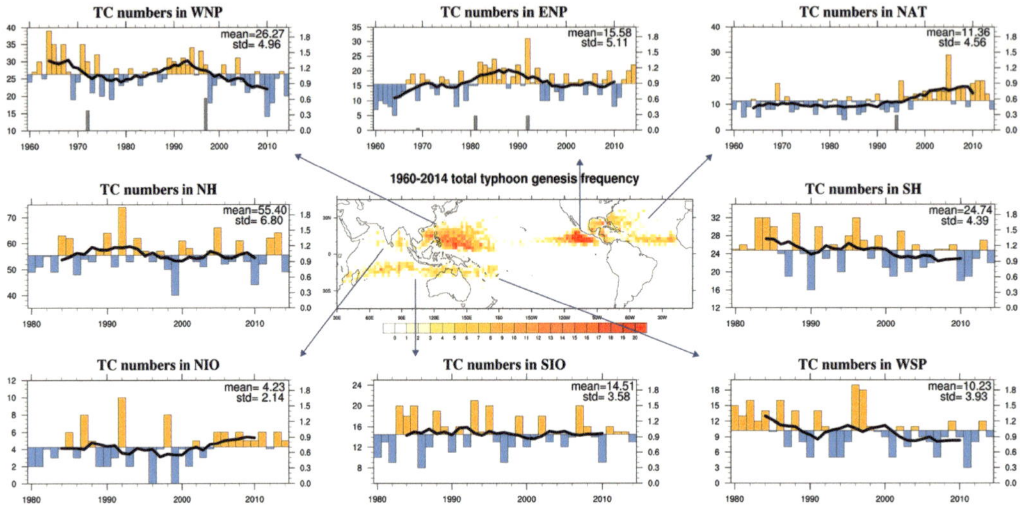

Figure 1. Time series of annual (filled bar) and 9-year running mean (black line) TC genesis number in various basins. The numbers in the upper corner indicate the mean and one standard deviation of the relevant to time series. The gray bar at the bottom shows the regime-shift index (RSI) [22] with 10-year cutoff length. Only the RSI exceeding 95% confidence is plotted. The central plot is the climatology of the spatial distribution of TC genesis density during 1960–2014.

all the TC scales (TS, TY, and ST) in the ENP increased simultaneously in the active period (1984–1993), however, the genesis number of the TY and ST exhibited a decrease in the inactive period in 1995–2004. The abrupt increase of TC activity in the North Atlantic (NAT) in middle 1990s was evident, that is, the TC genesis number exhibited a significant interdecadal increase after 1994/1995 [10, 14, 15]. In contrast to the ENP, the TC number in the NAT only experienced once abrupt change in 1994 (**Figure 1**). This abrupt change is primarily resulted from the number increase in TS and especially in ST, however, the contribution of TY is relatively weak (**Figure 2c**). It is clear that the abrupt increase of TC number in the ENP in the early 1980s is concurrent with the abrupt increase in the WNP. Whereas the abrupt decrease of TC number in the ENP in mid-1990s is synchronized with the abrupt increase in the NAT.

The time series of TC activity in the WSP shows that the TC genesis number was high in 1980–1990 and low throughout the 2000s. Whereas the total TC number decreases since 2000, the number of ST shows a gradual increase (**Figure 2d**). The decrease since 2000 was approximately synchronized with the abrupt decrease of TC number in the WNP in the late 1990s. The TC activity in the North Indian Ocean (NIO) was low during the 1990–2000, but was high, especially for the specific scale of TS, during the early 2000s (**Figure 2e**). Conversely, the TC activity in the South Indian Ocean (SIO) was high during middle 1980s to middle 1990s (**Figure 2f**). In contrast to North Hemisphere, the TC activity in South Hemisphere does not show significant abrupt change. Only the TC number in the WSP shows an abrupt decrease approximately in 2000 with a lower confidence level at 80%.

Overall, the total TC number in all the genesis regions in each year is approximately the same. Approximately 80 TCs forms each year in the global (**Figure 3a**). That is, the global TC number

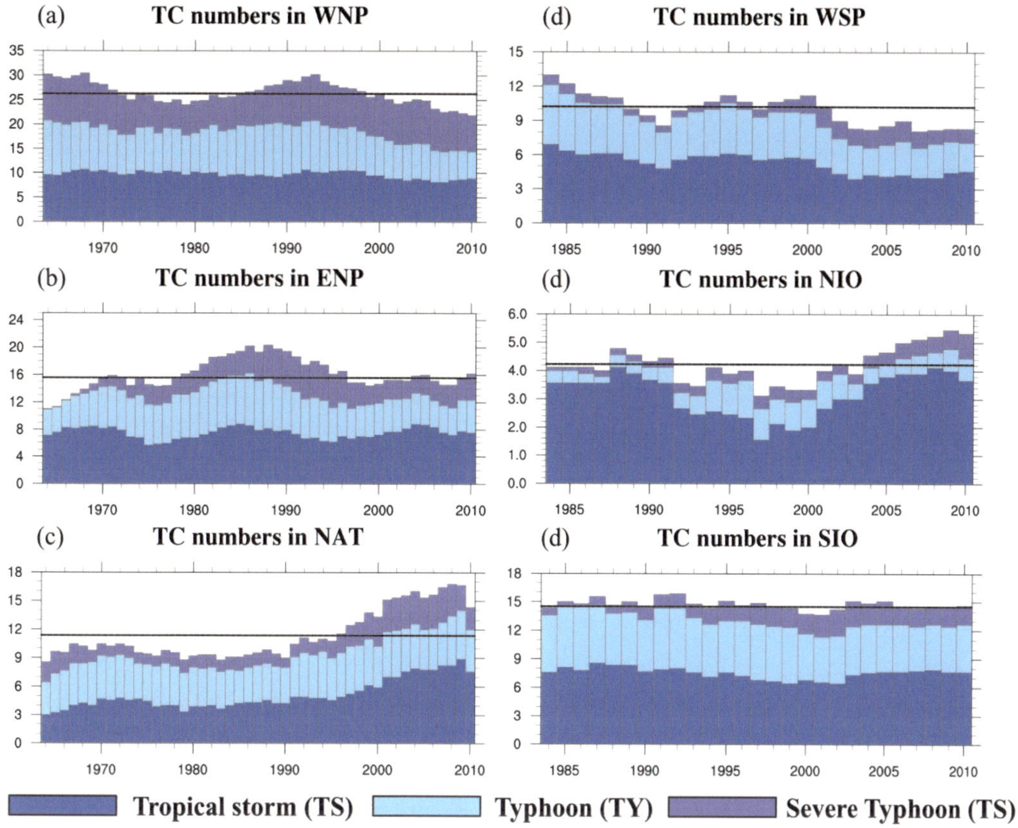

Figure 2. Time series of 9-year running mean (filled bar) of TC genesis number in various basins. Colorful fill bar represent the genesis number of different scale of TCs, including TS, TY and ST (**Table 1**). The horizontal dot line denotes the climatological mean of the TC genesis number.

does not exhibits interdecadal variation as identified in specific genesis region, such as in the WNP. The annual mean TC number in North Hemisphere also approximately remains the same since the late 1990s (**Figure 3b**) because that the decrease of TC number in the WNP is nearly compensated by the increase of TC number in the NAT (**Figure 2**). By contrast, the annual mean TC number in South Hemisphere shows an interdecadal decrease since the late 1990s, and the decrease is primarily resulted from the decline in specific scale of TY (**Figure 3c**). We note that the time series of TS and TY number for both Hemispheres are approximately opposite since the late 1990s. That is the TS and TY genesis number exhibits gradually increase and decrease respectively since the late 1990s. The interdecadal variation of the ST number in both Hemispheres is not so pronounced compared with that of TS and TY. The opposite relationship between the TS number and TY number is clearly seen in the ratio of specific scale of TC number to the total TC number: the ratio of TS and ST number to the total TC number in North Hemisphere and global exhibit gradually increase trend conversely, the ratio of TY number to the total TC number shows decrease trend since the late 1990s (**Figure 3d–f**).

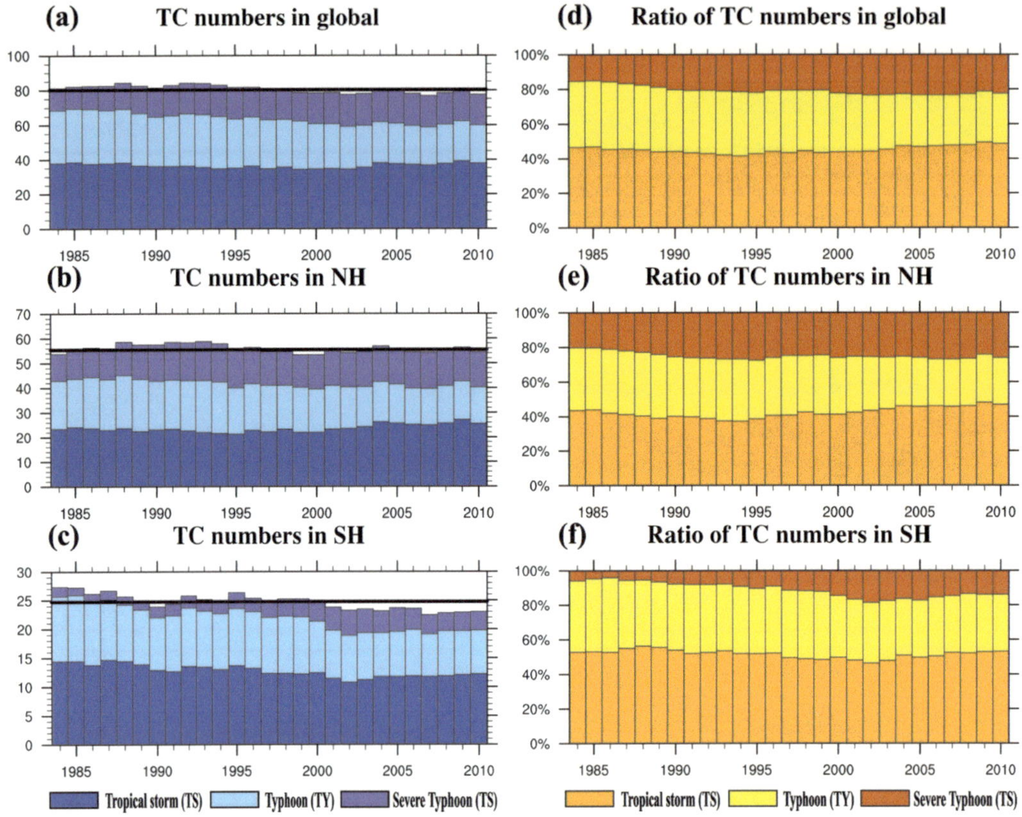

Figure 3. (Left) Same as in **Figure 2** except for the TC number in global, North and South Hemispheres. (Right) Same as in left panels except for the ratio of TC number of specific scale to the total TC number.

Based on the analysis of RSI in TC (TS + TY + ST) number, four significant abrupt changes of TC genesis number in various regions are identified during 1960–2014 (**Table 2**), including (i) an abrupt decrease (increase) in the WNP (ENP) in the early 1970s, (ii) an abrupt increase in the ENP and WNP in the early 1980s, (iii) an abrupt increase (decrease) in the NAT (ENP) in the middle 1990s, and (iv) an abrupt decrease in the WNP and WSP in the late 1990s. In the first abrupt change, the annual mean TC number in the WNP (ENP) decreases (increases) approximately 15% year^{-1} after the abrupt change in the late 1990s; In the second abrupt change, the TC number in the ENP increases approximately 40% year^{-1}; the TC number in the ENP (NAT) decreases (increases) approximately 36% year^{-1} in the third abrupt change; The last change is especially pronounced, the TC number in the WNP (WSP) decreases approximately 20% year^{-1}, and the TC number in the ENP (NAT) decreases (increases) approximately 15% (42%) year^{-1}. Whereas the TC number in the NIO also exhibits a remarkable increase (26% year^{-1}) after the mid-1990s, this increase is not significant because that the climatology of annual mean TC number is quite low (approximately 4), a small deviation of TC number could lead to a larger change.

Time of an abrupt change of TC activity		WNP Mean = 26.27 Std. = 4.96		ENP Mean = 15.58 Std. = 5.11		NAT Mean = 11.36 Std. = 4.56		NIO Mean = 4.23 Std. = 2.14		WSP Mean = 10.23 Std. = 3.77		SIO Mean = 14.40 Std. = 3.26	
1971	1961–1970	29.1	−4.1 (15.4%)	12.1	2.3 (14.8%)	9.7	−0.2 (1.8%)	–		–		–	
	1972–1981	25		14.4		9.5							
1982	1972–1981	25	2.4 (9%)	14.4	6.1 (39.2%)	9.5	−0.2 (1.8%)	–		–		–	
	1983–1992	27.4		20.5		9.3							
1994	1984–1993	28	−1.6 (6%)	19.9	−4.8 (30.8%)	9.8	3.3 (29%)	4.5	−1.1 (26%)	10.4	−0.1 (1%)	15.3	−0.9 (6.3%)
	1995–2004	26.4		15.1		13.1		3.4		10.3		14.4	
1998	1988–1997	29.7	−4.8 (18%)	17.6	−2.3 (14.8%)	10.8	4.8 (42.3%)	3.6	0.5 (11.8%)	10.7	−2.3 (22.5%)	15.2	0 (0%)
	1999–2008	24.9		15.3		15.6		4.1		8.4		15.2	

A regime shift index [22] was used to investigate an abrupt change of TC genesis number. Ten-years mean is used to indicate the time mean (background) state.

Table 2. Difference of the annual mean of TC number in various genesis regions after and prior the abrupt change of TC activity in 1971, 1982, 1994 and 1998.

4. Influence of climate regime shift on abrupt change of TC activity

In this section, we will demonstrate that the possible effect of a CRS on the abrupt change of TC genesis number. A CRS is featured by an abrupt transition from one quasi-steady climate state (or time mean state) to another quasi-steady state, and its transition period is relative shorter than the duration of each climate state. Using an analysis of experience orthogonal function (EOF) on SST, Yasunaga and Hanawa [23] reported that the North Hemisphere winter experienced three significant CRSs in 1971/1972, 1976/1977, and 1988/1989, during 1960–1997. The first CRS in 1971/1972 resembles the Central-Pacific or Modoki-El Niño SST anomaly [24–27], and the CRS in 1976/1977 and 1988/1989 are referred to a positive PDO or IPO pattern.

	PC1	PC2	AMO	PDO	CP	IPO
PC1	–					
PC2	−0.10	–				
AMO	−0.55	−0.54	–			
PDO	0.60	0.05	−0.09	–		
CP	0.25	−0.69	0.45	0.36	–	
IPO	0.73	−0.24	−0.03	0.85	0.58	–

A 9-year running mean is applied to the index before the calculation. The PCs are the principle component of SVDs. It indicates the SVD1 (SVD2) is closely related with the IPO (CP), and both SVD1 and SVD2 are correlated with AMO. The AMO, PDO and IPO indices are obtained from the website at https://www.esrl.noaa.gov/psd/data/timeseries/AMO/, https://www.esrl.noaa.gov/psd/gcos_wgsp/Timeseries/PDO/, https://www.esrl.noaa.gov/psd/data/timeseries/IPOTPI/ respectively. The CP index is calculated by the previous study [28].

Table 3. Correlation coefficient table among the indices of SVDs, AMO, IPO, PDO and CP during 1960–2010.

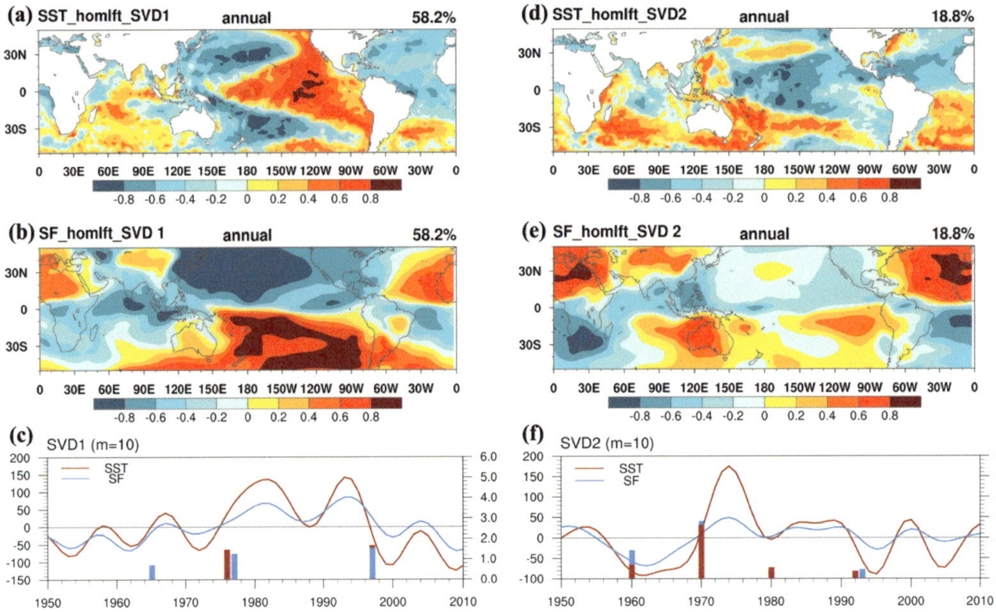

Figure 4. (Left) The leading SVD pattern of (a) SST and (b) 850 hPa stream function fields and (c) the corresponding principle components (PCs). The bars in (c) indicate the PCs experienced a significant abrupt change detected by regime shift index [22]. (Right) Same as in the left panels but for the second SVD mode.

These three CRSs are further identified by SVD analysis by paring the global SST and low-level circulation (850 hPa stream function) [16], indicating these CRSs are regime shifts of the air-sea coupled climate system. The SVD analysis reveals that leading SVD mode, with spatial pattern resembling the IPO (**Table 3**), showed a significant abrupt change in the 1976/1977 and 1996/1997 (**Figure 4a**); nevertheless, the second SVD mode, characterized by a CP La Niña-like [28] or Pacific Meridional Mode (PMM)-like [29, 30] SSTA in the Pacific (i.e., a southwest-northeast tilted negative SSTA in the central-eastern North Pacific, as well as a cooling in the tropical Atlantic), exhibited an abrupt change in the 1970/1971. A comparison between **Table 2** and the CRSs indicates that the three abrupt changes of TC activity in the Pacific basin, including the early 1970s, early 1980s and late 1990s, were approximately synchronized with the CRSs in the 1971/1972, 1976/1977 and 1996/1997 respectively (**Figure 5**). Whereas the abrupt TC increase in the NAT in the 1993/1994 was not corresponding to a specific CRS, it was concurrent with the interdecadal change of AMO.

4.1. Effect of CP-like SST on the abrupt change of TC activity in the early 1970s

The CRS in the early 1970s is primarily determined by the second SVD, which is characterized by a CP La Niña-like SSTA (**Figure 4d**). **Figure 5** shows that the CRS in the early 1970s is concurrent with the abrupt decrease/increase of TC number in the WNP/ENP in the early 1970s, suggesting that both are related. Because the TC activity is substantially affected by the large-scale factors (e.g., low-level vorticity, vertical wind shear and mid-atmosphere relative humidity), the effect of CRS on the abrupt change of TC activity can be understood by regressing these large-scale factors on the SVD. Because the first two SVDs are closely related

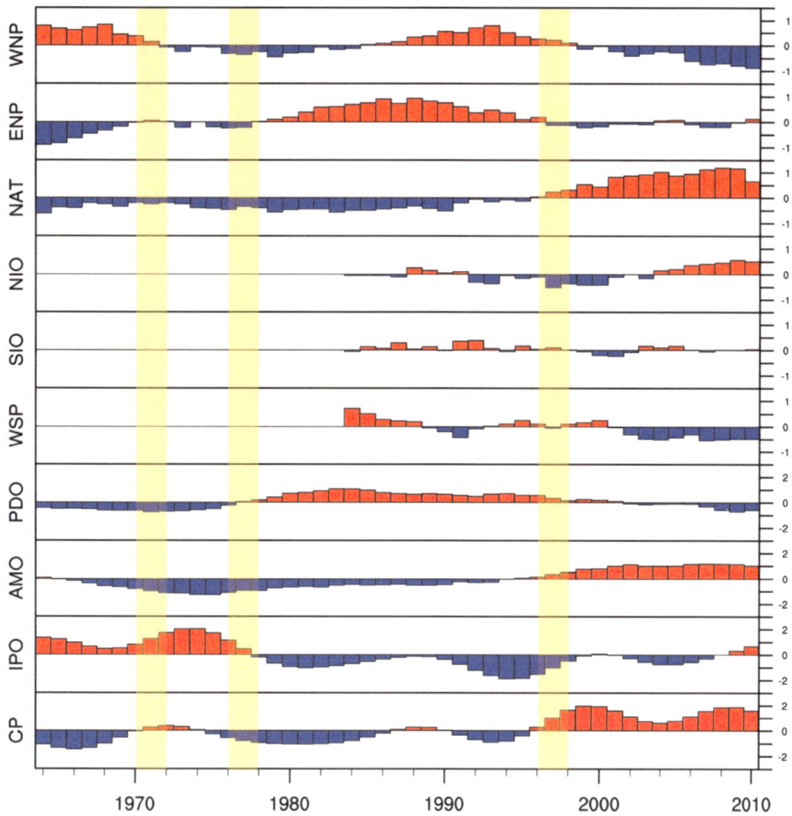

Figure 5. The bars indicate the normalized TC genesis number in various genesis regions and PDO, AMO, IPO and CP indices. The vertical yellow shaded indicates that the climate regime shift in 1971/1972, 1976/1977 and 1996/1997. The climate regime shift is identified by the regime shift index of leading SVDs by coupling the SST and 850 hPa stream function (**Figure 4c** and **f**).

with the AMO (**Table 3**), the partial regression by removing the effect of AMO is applied to investigate the net effect of SVD on the large-scale conditions. It shows that the second SVD is associated with a pair of cyclonic anomalies in the tropical eastern Pacific, as well as an anticyclonic circulation anomaly in the WNP and WSP (**Figure 6d** marked with H and L, respectively). This anticyclonic circulation creates a low-level easterly anomaly and a positive zonal wind shear in the WNP (**Figure 6e**), which provides unfavorable large-scale conditions for TC activity. Conversely, the cyclonic circulation creates a favorable large-scale environment for TC genesis in the ENP. In contrast to the dynamic factors, the regression of thermodynamics (700 hPa relative humid) does not exhibit significant signal in the WNP and ENP (**Figure 6f**). That indicates the CP La Niña-like SSTA associated dynamic conditions determine the abrupt decrease (increase) of TC genesis number in the WNP (ENP) in the early 1970s.

4.2. Effect of IPO-like SST on the abrupt change of TC activity in the early 1980s

The CRS in the early 1976/1977 is determined by the leading SVD, featured by a positive IPO SSTA. The TC activity-related large-scale factors regressed on the leading SVD reveals that the

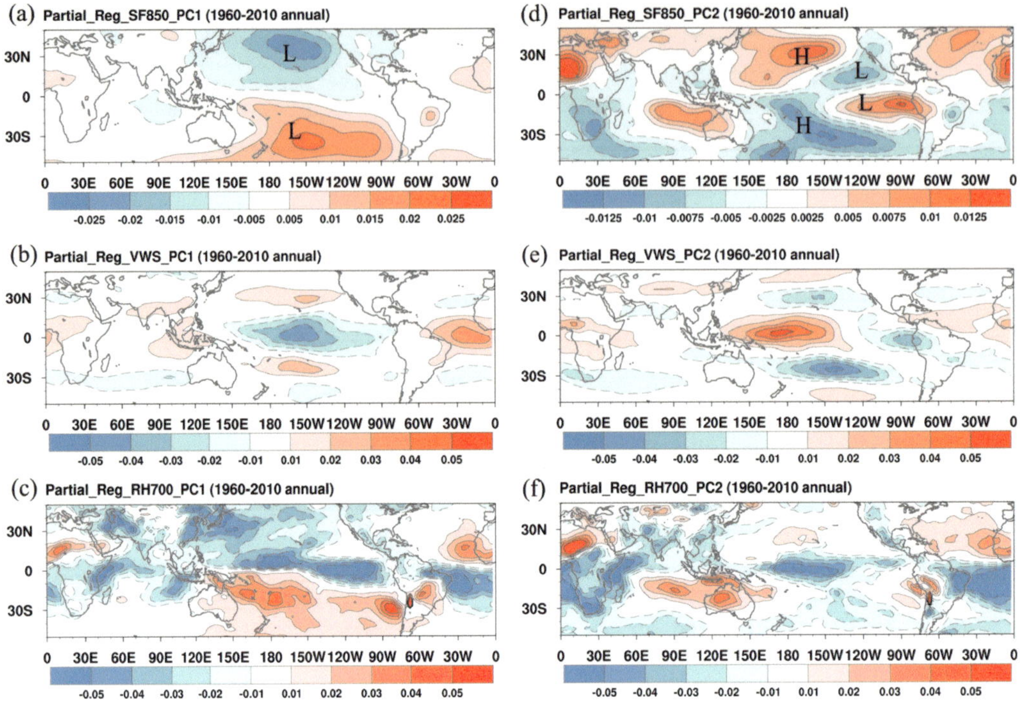

Figure 6. The partial regression (with AMO index fixed) of TC activity associated large-scale dynamic and thermal conditions on the first SVD (IPO-like SST): (a) 850 hPa stream function (b) vertical zonal wind shear, and (c) 700 hPa relative humidity. (d–f) Same as in (a–c) except for the second SVD (CP-like SST).

first SVD (i.e., IPO or PDO) was accompanied by a pair of basin–scale cyclonic anomaly, a Matsuno-Gill response [31, 32] to the positive SSTA in the tropical eastern Pacific (**Figure 6a**). The pair of cyclonic anomaly creates a negative vertical zonal wind shear anomaly in the tropical eastern Pacific (**Figure 6b**), which provides a favorable dynamical condition for TC genesis. There has no evidence indicates that the mid-level relative humid contributes to the abrupt increase of TC genesis number in the WNP and ENP in the early 1980s. In contrast to the dynamic factors, the positive IPO phase is associated with a negative 700 hPa relative humidity anomaly in the WNP (**Figure 6c**), which is unfavorable for TC genesis. Because the effect of dynamics on enhancing the TC activity in the WNP partially offsets by the effect of thermodynamics, the magnitude of abrupt increase of TC number in the WNP in the early 1980s is much weaker than that in the ENP. Again, the IPO-related large-scale change in the dynamics determines the abrupt increase of TC number in the ENP and WNP in the early 1980s.

4.3. Combined effect of IPO and AMO on the abrupt change of TC activity in the late 1990s

The CRS in 1996/1997 is also primarily determined by the leading SVD except that the amplitude is negative. That is the CRS-induced time mean state change exhibits a negative IPO-like (or Mega La Niña-like) SSTA pattern, which is accompanied by a pair of anticyclone and

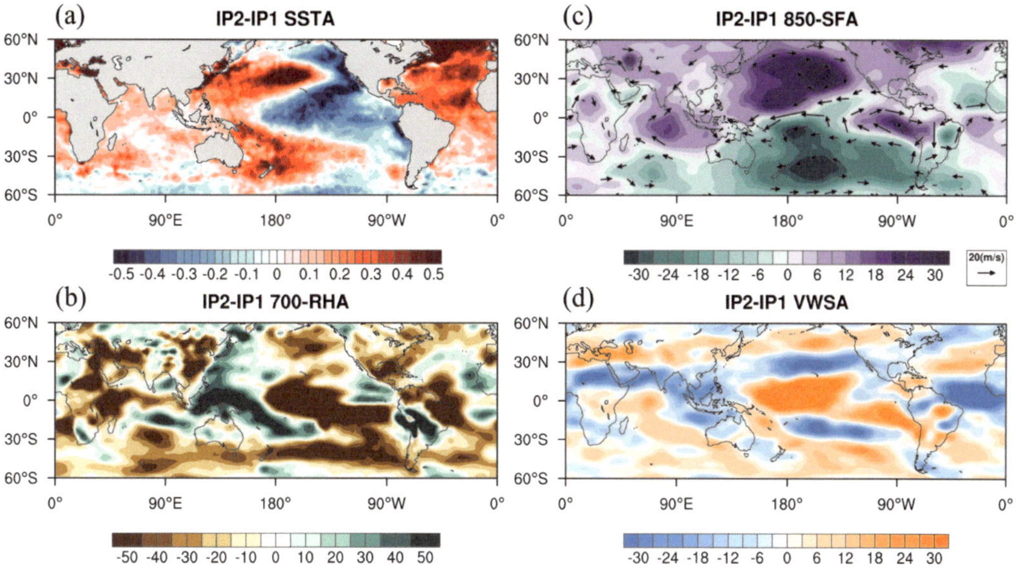

Figure 7. The time mean-state difference relevant to the climate regime shift in 1996/1997 ([IP2: 1999–2008 minus IP1: 1989–1997). (a) SST (°C), (b) 700 hPa relative humidity (%), (c) 850 hPa stream function (10^{-6} m^2 s^{-1}), (d) vertical zonal wind shear (m s^{-1}) (U_{200}–U_{850}).

cyclone anomalies in the western Pacific and eastern Pacific respectively (**Figure 7**). The anticyclone associated positive vertical zonal wind shear anomaly in the western Pacific is not favorable for the TC activity in the WNP and WSP. The influence of the basin-wide CRS in 1996/1997 on the TC activity in the late 1990s could be interpreted as outlined in follows: the global SST experienced a significant regime shift approximately in 1996/1997. Prior to this climate regime shift, a positive SSTA appeared in the equatorial central-eastern Pacific as well as a negative SSTA appeared in the tropical Atlantic and western Pacific. In response to this positive SSTA, a Matsuno-Gill-type [31, 32] of low-level cyclonic circulation anomaly occurred in the WNP. This large-scale anomalous cyclone, together with reduced vertical zonal wind shear, favored TC genesis, which explains why TC number in the WNP were above normal before the CRS. However, these anomalous thermodynamic and dynamic conditions reversed after 1996/1997. That is, the SSTA in the equatorial central-eastern Pacific turned from positive to negative, and vice versa in the equatorial western Pacific, after the CRS in 1996/1997. In response to the phase change of the SSTA in the central-eastern equatorial Pacific, an anticyclone anomaly was observed in the WNP/WSP after 1996/1997. Meanwhile, the anticyclone associated low-level easterly wind anomaly creates a positive vertical zonal wind shear anomaly in the western Pacific. Both of these changes in the background state suppressed TC activity in the WNP and WSP.

A comparison between the time mean state change prior and after the CTS in 1996/1997 (**Figure 7**) and the leading SVD (**Figure 4a** and **b**) reveals that a pair of cyclonic anomaly in the ENP (**Figure 7**) is not detected in **Figure 4a** and **b**. This distinction is primarily caused by the effect of the interdecadal change of AMO, which terminated phase in 1994/1995 and was approximately

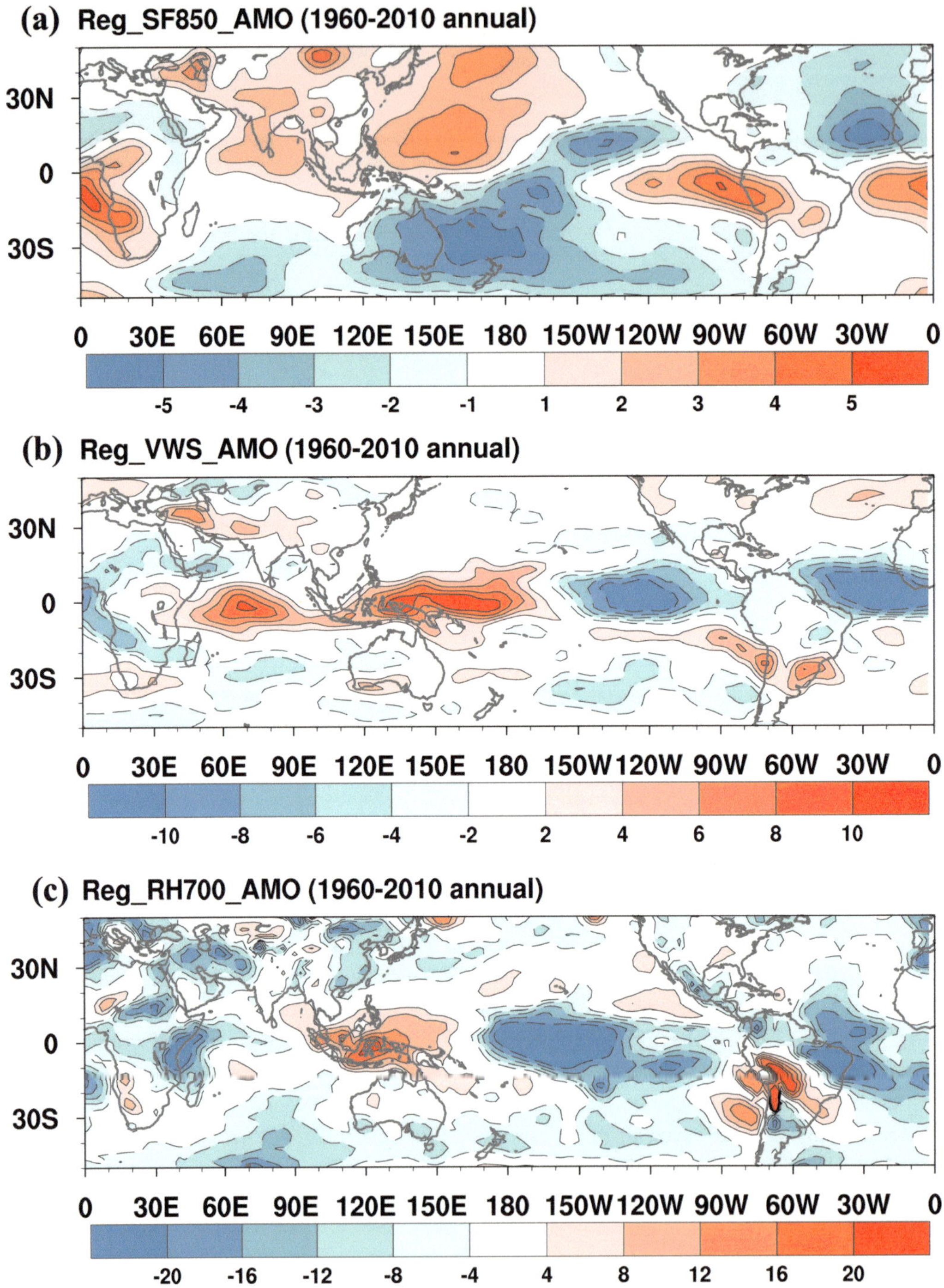

Figure 8. Same as in **Figure 7** except for the AMO index.

http://dx.doi.org/10.5772/intechopen.74947

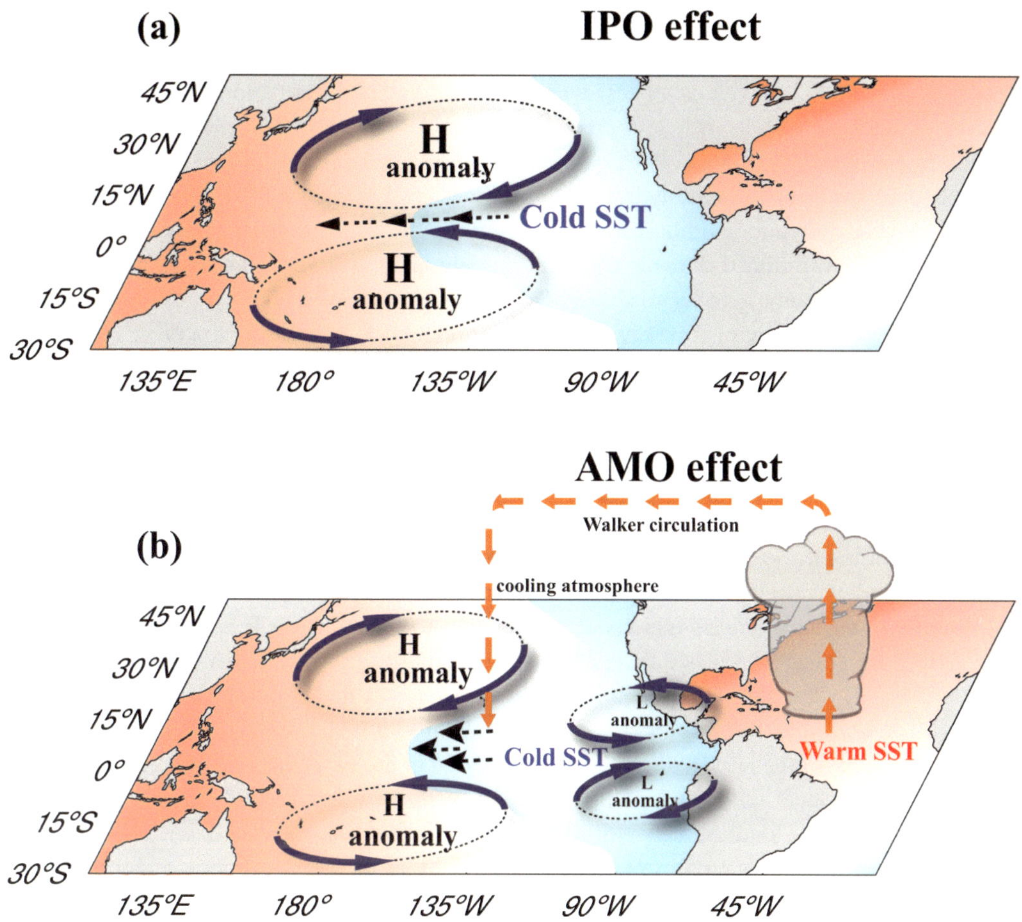

Figure 9. Schematic diagram illustrates how the IPO and AMO affects the abrupt change of TC activity in late 1990s.

synchronized with the CRS in 1996/1997 in the view of interdecadal timescale. The regression of 850 hPa stream function on the AMO reveals a pair of cyclone anomaly in the ENP and a pair of anticyclone anomaly in the western Pacific (**Figure 8a**). That indicates the effect of negative IPO phase on suppressing the TC activity in the western Pacific was fuehrer enhanced by the remote impact of a positive AMO associated positive SSTA in the tropical Atlantic (**Figure 8b–c**). The AMO not only exerts a remarkable remote impact on the TC activity in the Pacific, it associated local change in thermodynamic also directly contributes to the abrupt increase of TC number in the Atlantic in 1993/1994 [10, 14, 15].

The combined effect of IPO and AMO on the TC activity in the Pacific in the late 1990s is presented in the schematic diagram (**Figure 9**) and described as followings: the negative IPO associated Mega La Niña SST anomaly forces a pair of basin-scale anomaly in the Pacific, in which the anticyclone associated negative vorticity and positive vertical zonal wind shear anomalies may substantially suppress the TC activity in the WNP and WSP (**Figure 9a**). Meanwhile, the AMO associated positive SSTA in tropical Atlantic drives a pair cyclonic circulation anomaly in the ENP, as well as an east-west overturning circulation anomaly in the Atlantic-Pacific, with subsidence branch in the equatorial central Pacific (**Figure 9b**). The downward motion in the equatorial central Pacific produces a low-level easterly anomaly and a positive vertical zonal wind shear anomaly in the tropical western Pacific, which provides an unfavorable large-scale dynamic condition for TC activity in the WNP and WSP. Additionally, the downward motion induced diabetic cooling anomaly generates a Rossby wave-like response of an anticyclone anomaly in its northwest (southwest) in the WNP (WSP), which in term the associated negative vorticity anomaly tends to suppress the TC activity in the WNP (WSP).

5. Conclusion

The possible impact of basin-scale CRS on the abrupt change of TC activity in various genesis regions, including the Pacific, Indian Ocean and Atlantic Ocean was investigated in this report. Here, a RSI index is use to detect a climate regime shift and an abrupt change of annual mean TC number in various genesis regions. It reveals that the worldwide TC activity experienced four significant abrupt changes during 1960–2014, including (i) an abrupt increase/decrease in the ENP/WNP in the early 1970s, (ii) an abrupt increase in the ENP and WNP in the early 1980s, (iii) an abrupt increase in the NAT/ENP in the middle 1990s, and (iv) an abrupt decrease in the WNP/WSP in the late 1990s. The abrupt change in the early 1970s, early 1980s and late 1990s is closely related with the CRS in 1971/1972, 1976/1977 and 1996/1997 respectively. The other abrupt change identified in North Atlantic in 1993/1994, is related with the decadal variation of AMO. That is an abrupt change of TC activity in specific genesis region is generally accompanied by an abrupt TC activity change in the other region.

Observation reveals that the abrupt change of TC number in various genesis regions is closely related with a basin-scale CRS or low frequency variability of the coupled climate system. The SVD analysis reveals that the CRS in 1976/1977 and 1996/1997 is primarily determined by the leading mode (i.e., IPO-like SST), whereas, the CRS in 1971/1972 is primarily dominated by the second SVD (i.e., CP La Niña-like SST). Among the four abrupt changes of TC activity, the last abrupt change in the late 1990s is especially pronounced because that the IPO and AMO joint together to contribute to the interdecadal change of TC activity in the Pacific basin. We demonstrated that the CRS-induced large scale oceanic and atmospheric background changes may lead to an abrupt change in TC activity in various genesis regions. The effect of CRS on the abrupt change of TC activity is further supported by the regression analysis of the TC activity-related large-scale factors on the first two SVDs.

It has been identified that an abrupt change occurred in TC activity in the various genesis regions of the Pacific during the late 1990s. The interdecadal decrease in TC activity in the WNP during the late 1990s had been reported widely in recent studies [8, 12, 13, 33]. However, in this chapter, a new interpretation of this interdecadal change has been proposed. Here, we

http://dx.doi.org/10.5772/intechopen.74947

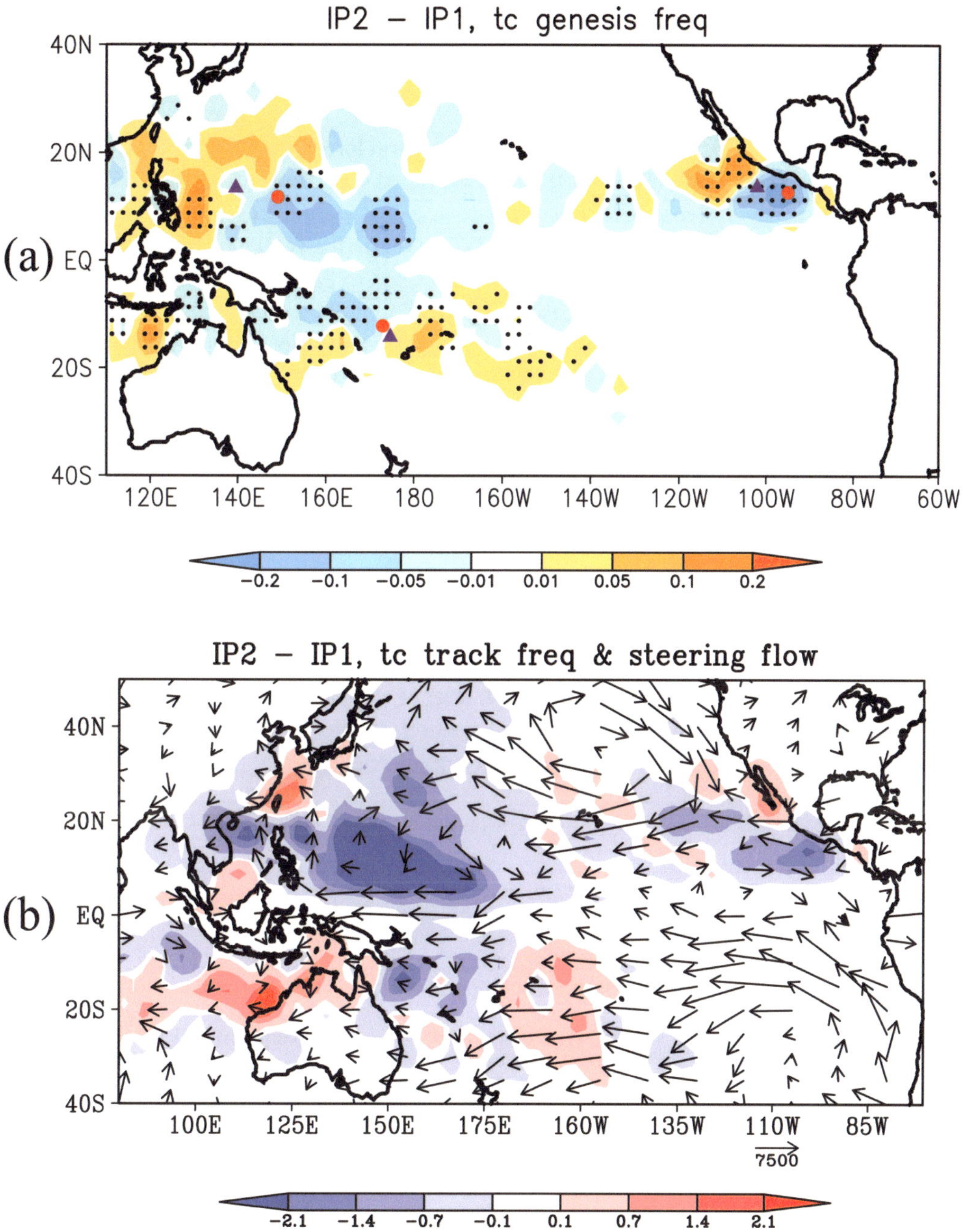

Figure 10. Difference of (a) TC genesis density and (b) TC track density relevant to the climate regime shift in 1996/1997 between the IP1 (1989–1997) and IP2 (1999–2008). The symbols "●" and " ▲" denotes the mean TC genesis position during IP1 and IP2, respectively. The dots in (a) indicate that the change in TC genesis frequency was at 5% significance level. The vectors in (b) indicate the steering flows that are defined as the vertical integrated horizontal winds from 850 to 300 hPa.

assumed that the abrupt change in TC activity in the WNP during the late 1990s was part of the phenomenon of basin-scale change in TC activity in the Pacific basin. In addition the TC genesis number, the CRS in the late 1990s had also resulted in a significant change in the mean TC genesis location in the WNP and ENP. The mean TC genesis location in WNP shift westward significantly from 149°E to 140°E, and the mean location in ENP shifts from approximately from 95°W to 102°W (**Figure 10a**). This westward shift in the ENP leads to a change in TC genesis frequency that resembles the response to global warming [34]. Of particular note is that the westward shift of the mean location of TC genesis in the WNP was accompanied by an increase in TC activity in the western WNP and southeastern part of the South China Sea (e.g., [35]). Studies had shown that the increase in TC activity in the western WNP could lead to a significant increase in TC landfalls in Taiwan [13, 36]. Irrespective of the increase in TC genesis frequency, our study showed that the modulation of the CRS-induced circulation on the steering flow (i.e.., westward extension and northward curving near the Philippine Sea) also contributed substantially to the increase in TC landfalls in Taiwan after late 1990s (**Figure 10b**).

Acknowledgements

We thank Mr. J.-K. Wu and C.-C. Chang to create the figures and tables used in this chapter. The observational atmospheric and oceanic data used in this study was from the NOAA's Earth System Research Laboratory and NCEP/NCAR. Data of tropical cyclone is from IBTrACS. This study was supported by the Ministry of Science and Technology, Taiwan, under grant MOST 105-2621-M-865-001, and MOST-105-2111-M-845-002 and MOST-106-2111-M-845-001.

Author details

Chi-Cherng Hong[1]* and Yi-Kai Wu[1,2]

*Address all correspondence to: cchong@utaipei.edu.tw

1 Department of Earth and Life Sciences, University of Taipei, Taipei, Taiwan

2 Department of Earth Sciences, National Taiwan Normal University, Taipei, Taiwan

References

[1] Chan J. Tropical cyclone activity over the western North Pacific associated with El Niño and La Niña events. Journal of Climate. 2000;**13**:2960-2972

[2] Chia H, Ropelewski C. The interannual variability in the genesis location of tropical cyclones in the northwest Pacific. Journal of Climate. 2002;**15**:2934-2944

[3] Wang B, Chan J. How strong ENSO events affect tropical storm activity over the western North Pacific. Journal of Climate. 2002;**15**:1643-1658

[4] Matsuura T, Yumoto M, Iizuka S. A mechanism of interdecadal variability of tropical cyclone activity over the western North Pacific. Climate Dynamics. 2003;**21**:105-117. DOI: 10.1007/s00382-003-0327-3

[5] Camargo S, Sobel A. Western North Pacific tropical cyclone intensity and ENSO. Journal of Climate. 2005;**18**:2996-3006

[6] Chan J. Interannual and interdecadal variations of tropical cyclone activity over the western North Pacific. Meteorology and Atmospheric Physics. 2005;**89**:143-152. DOI: 10.1007/s00703-005-0126-y

[7] Li Z, Yu W, Li T, Murty V, Tangang F. Bimodal character of cyclone climatology in Bay of Bengal modulated by monsoon seasonal cycle. Journal of Climate. 2013;**26**:1033-1046

[8] Liu K, Chan J. Inactive period of western North Pacific tropical cyclone activity in 1998–2011. Journal of Climate. 2013;**26**:2614-2630

[9] Lander M, Guard C. A look a global tropical cyclone activity during 1995: Contrasting high Atlantic activity with low activity in other basins. Monthly Weather Review. 1998; **126**:1163-1173

[10] Saunders M, Lea A. Large contribution of sea surface warming to recent increase in Atlantic hurricane activity. Nature. 2008;**451**:557-561. DOI: 10.1038/nature06423

[11] Lupo A, Latham T, Magill T, Clark J, Melick C, Market P. The interannual variability of hurricane activity in the Atlantic and East Pacific Regions. National Weather Digest. 2008; **32**:119-135

[12] Choi Y, Ha K, Ho C, Chung C. Interdecadal change in typhoon genesis condition over the western North Pacific. Climate Dynamics. 2015;**45**(11–12):3243-3255

[13] He H, Yang J, Gong D, Mao R, Wang Y, Gao M. Decadal changes in tropical cyclone activity over the western North Pacific in the late 1990s. Climate Dynamics. 2015;**45**(11–12): 3317-3329

[14] Goldenberg S, Landsea C, Mestas-Nuñez A, Gray W. The recent increase in Atlantic hurricane activity: Causes and implications. Science. 2001;**293**:474. DOI: 10.1126/science.1060040

[15] Lupo A. The interannual and interdecadal variability in hurricane activity. In: Lupo A, editor. Recent Hurricane Research—Climate, Dynamics, and Societal Impacts. Rijeka: InTech; 2011, ISBN: 978-953-307-238-8. Available from: http://www.intechopen.com/books/recent-hurricane-research-climate-dynamics-and-societal-impacts/the-interannual-and-interdecadal-variability-in-hurricane-activity

[16] Hong C, Wu Y, Li T, Chang C. The climate regime shift in Pacific during 1996/1997. Climate Dynamics. 2013;**43**:435-446. DOI: 10.1007/s00382-013- 1867-9

[17] Hong C, Wu Y, Li T. Influence of climate regime shift on the interdecadal change in tropical cyclone activity over the Pacific Basin during the middle to late 1990s. Climate Dynamics. 2016;**47**(7):2587-2600. DOI: 10.1007/s00382-016-2986-x

[18] Knapp K, Kruk M, Levinson D, Diamond H, Neumann C. The international best track archive for climate stewardship (IBTrACS): Unifying tropical cyclone best track data. Bulletin of the American Meteorological Society. 2010;**91**:363-376. DOI: 10.1175/2009BAMS2755.1

[19] Kalnay E et al. The NCEP/NCAR 40-year reanalysis project. Bulletin of the American Meteorological Society. 1996;**77**:437-471

[20] Rayner N, Parker D, Horton E, Folland C, Alexander L, Rowell D, Kent E, Kaplan A. Global analyses of sea surface temperature, sea ice, and night marine air temperature since the late nineteenth century. Journal of Geophysical Research. 2003;**108**:4407. DOI: 10.1029/2002JD002670

[21] Simpson R. The hurricane disaster potential scale. Weatherwise. 1974;**27**:169-186

[22] Rodionov S. A sequential algorithm for testing climate regime shifts. Geophysical Research Letters. 2004;**31**:L09204. DOI: 10.1029/2004GL019448

[23] Yasunaga S, Hanawa K. Regime shifts found in the northern hemisphere SST field. Journal of the Meteorological Society of Japan. 2002;**80**:119-135

[24] Ashok K, Behera S, Rao A, Weng H, Yamagata T. El Niño Modoki and its teleconnection. Journal of Geophysical Research. 2007;**112**:C11007. DOI: 10.1029/2006JC003798

[25] Kao H, Yu J. Contrasting Eastern-Pacific and Central-Pacific types of ENSO. Journal of Climate. 2009;**22**:615-632

[26] Kug J, Jin F, An S. Two types of El Niño events: Cold tongue El Niño and warm pool El Niño. Journal of Climate. 2009;**22**:1499-1515

[27] Lee T, McPhaden M. Increasing intensity of El Niño in the central-equatorial Pacific. Geophysical Research Letters. 2010;**37**:L14603. DOI: 10.1029/2010GL044007

[28] Yu J, Lu M, Kim S. A change in the relationship between tropical central Pacific SST variability and the extratropical atmosphere around 1990. Environmental Research Letters. 2012;**7**:034025

[29] Chiang J, Vimont D. Analogous Pacific and Atlantic meridional modes of tropical atmosphere–ocean variability. Journal of Climate. 2004;**17**:4143-4158

[30] Chang P, Zhang L, Saravanan R, Vimont D, Chiang J, Ji L, Seidel H, Tippett M. Pacific meridional mode and El Niño—Southern oscillation. Geophysical Research Letters. 2007; **34**:L16608h

[31] Matsuno T. Quasi-geostrophic motions in the equatorial area. Journal of the Meteorological Society of Japan. 1966;**44**:25-43

[32] Gill A. Some simple solutions for heat-induced tropical circulation. Quarterly Journal of the Royal Meteorological Society. 1980;**106**:447-462

[33] Hsu P, Chu P, Murakami H, Zhao X. An abrupt decrease in the late-season typhoon activity over the Western North Pacific. Journal of Climate. 2014;**27**:4296-4312. DOI: 10.1175/JCLI-D-13-00417.1

[34] Murakami H, Wang B, Li T, Kitoh A. Projected increase in tropical cyclones near Hawaii. Nature Climate Change. 2013;**3**(8):749-754

[35] Yang L, Du Y, Xie S, Wang D. An interdecadal change of tropical cyclone activity in the South China Sea in the early 1990s. Chinese Journal of Oceanology and Limnology. 2012; **30**:953-959

[36] Tu J, Chou C, Huang P, Huang R. An abrupt increase of intense typhoons over the western North Pacific in early summer. Environmental Research Letters. 2011;**6**:034013. DOI: 10.1088 /1748-9326/6/3/034013

Chapter 3

The Highest Geomagnetic Storms of the Solar Cycle Observed at Ground Level

Carlos E. Navia, Marcel N. de Oliveira and
Carlos R. A. Augusto

Additional information is available at the end of the chapter

http://dx.doi.org/10.5772/intechopen.75688

Abstract

We report two ground-level observations, of geomagnetic storms of different origins; they are among the highest geomagnetic storms, in the solar Cycle 24. The first is St. Patrick's Day storm on March 17, 2015, originated by the impact on Earth's atmosphere of coronal mass ejections (CMEs), the storm reaching the condition of G4 (severe) level, in the NOAA geomagnetic scale. The second included the major geomagnetic storm whose origin is attributed to the interaction with the Earth of a High-Speed Stream (HSS) ahead of a positive polarity coronal hole on October 7, 2015. This storm reached the condition G3 (strong) level. We give emphasis to observations detected by the New-Tupi muon telescopes, located at sea level in Brazil (22.53° S, 43.13° W). We present a study of these observations in correlation with observations reported by multipoint space-based measurements, such as the ACE at Lagrange Point L1 and the geostationary GOES weather satellite, including two global geomagnetic indices and several ground-based detectors. Some considerations on the influence of these geomagnetic storms in the Earth weather are reported.

Keywords: solar physics, geomagnetic storms, space weather, particle detectors

1. Introduction

The previous solar cycle, Solar Cycle 23, peaked in 2000–2002 with many furious solar storms. However, the current solar cycle 24 has not been good to observe geomagnetic storms. Years of weak solar activity have produced few widespread geomagnetic storms [1]. Since its inception, the current solar cycle 24 shows anomalies. March 2008 marked the end of cycle 23 and the beginning of cycle 24 [2]. This means cycle 23 was 11.75 years, higher than the average value of

11 years. In addition, until the second semester of 2010, the sun was in a low level of activity, in an anomalous extended period of minimal solar activity [3].

The current solar cycle 24 had a double maximum of activity [4, 5]; the first happened in late 2013 and another in 2015. In both cases, the solar activity has been relatively low. Now, December 2017, the cycle 24 already is in a declining phase. According to Pesnell from NASA Goddard Space Flight Center, there are similarities between the current cycle 24 and the solar cycle 14, which took place between February 1902 and August 1913 and experienced a double maximum and similar intensity. In addition, the last prediction on Solar Cycle 25 is expected to be the smallest in over 300 years. Suggesting the end of the Modern Warm Period, plummeting Earth into a period of global cooling, which could be like the Mini Ice Age that occurred during the last grand minimum of solar activity, known as the Maunder Minimum [6]. So far, as mentioned earlier, the activity of the solar cycle 24 was weak when compared, for instance, with its predecessor, the cycle 23. The number of geomagnetic storms [7, 8], triggered in each cycle can be used as a comparison between cycles. We have compiled the top 50 geomagnetic storms (using the NOAA weather scales) [9], that is, the 50 strongest geomagnetic storms from Solar Cycle 23 and (so far) on the cycle 24. **Figure 1** show the result of this comparison, where the intensity of a geomagnetic storm is characterized by two indices. The Kp index that indicates the disturbance of the horizontal geomagnetic field and ranging from 0 to 9, a value equal or above 5 indicates the condition of a geomagnetic storm and the G index (NOAA scale for geomagnetic storms), ranging from 1 to 4, G = 1 is equivalent to Kp = 5, and so on.

Following **Figure 1** we can see that the majority (top 50) of the geomagnetic storms on cycle 23 were of type G4 (severe) or Kp = 8, with 22 events. This contrasts with only 6 events (so far) in this category on the cycle 24. In addition, the highest geomagnetic storms, those cataloged as G5 (extreme) or k = 9 on cycle 23, was 13, and none observed in cycle 24. So far, the majority of the top 50 geomagnetic storms on cycle 24 were of type G2 (moderate) or Kp = 6, with 28 events.

The aim of this chapter is to present a global study of the effects on Earth of the highest (so far) geomagnetic storms on the current cycle 24, producing the so-called Forbush decrease (FD) [10–12]. They are defined as sudden decreases of the recorded galactic cosmic ray (GCRs) intensity detected

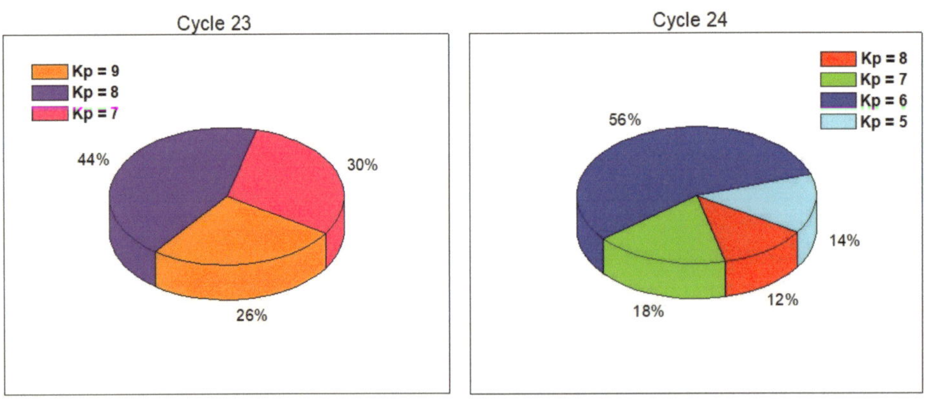

Figure 1. The top 50 geomagnetic storms classified according to Kp index, for Solar Cycle 23 (left panel) and (so far) for the cycle 24 (right panel).

at ground level, followed by a gradual nearly exponential recovery time of several days. The FD are known since long and form an interesting phenomenon. Geomagnetic storms and their ground-level effects can play an important role to understand the global aspects of the weather conditions and is a very active area of research.

The magnitude of FDs observed by a particular detector depends on several factors: the size of the CME, the strength of the magnetic fields in the CME, the proximity of the CME to the Earth, and the location of the detector on the Earth [13]. Initially, this study was limited to three events but a full study is in progress. We compare neutron monitor data and especially the data from the New-Tupi muon telescope with multipoint space-based measurements of the interplanetary space (e.g. ACE/SWEPAM and ACE/MAG and the geostationary GOES weather satellite) and several global geomagnetic indices.

The classification of geomagnetic storms is in two categories: the non-recurrent because the origin is due to the passage of interplanetary counterparts of a coronal mass ejection by Earth. The FD triggered by these geomagnetic storms has a decrease that typically lasts for less than 1 day while the recovery phase may last for several days.

There are also the recurrent geomagnetic storms, whose origins are linked with the interaction with the Earth's magnetic field of a coronal high-speed stream (HSS) [14]. These events dominate near the minimum of the 11-year solar activity cycle. They are recurrent phenomena [15], because, under certain conditions, i.e., stable coronal holes, the HSS can reappear after 27 days, this time is due to Sun rotation period. Even so, they can also appear on and near solar maximum. However, in this case they are less persistent [14]. They are also very important for driving activity in the Earth's magnetosphere, but they are in most cases in minor intensity than the geomagnetic storm originated by CMEs.

The first significant geomagnetic storm of solar cycle 24 was on February 18, 2011. This was the result of the coronal mass ejections ejected on February 14 and 15, 2011. This was also the first significant FD observed from neutron monitors around the world. While the highest geomagnetic storm (so far) was only on March 17, 2015, the so-called St. Patrick Day storm, reaching the condition of a G4 (severe) class, the analysis of this event, considered between the four highest geomagnetic storms of cycle 24, constitute the main core of this article. However, we also have included an analysis of the highest geomagnetic storm whose origin was the interaction with the Earth of a coronal high-speed stream on October 7, 2015. We give emphasis to the observation of the effect at ground level of these three geomagnetic storms through of their FD signatures in different detectors, one of them within the South Atlantic Anomaly (SAA). In this book, we report details of these observations.

2. The New-Tupi detector

The New-Tupi experiment is devoted to the study of cosmic rays and diverse transient events and space weather phenomena in Earth's interplanetary environment. It is an Earth-based apparatus located in the Physics Institute of the Universidade Federal Fluminense, Niteroi, Rio de Janeiro, Brazil (22.53° S, 43.13° W).

The New-Tupi detector is sensitive to primary particles with energies above the pion production threshold (particles with energies around and above GeV scales). The collisions of primary particles, such as the cosmic ray particles or solar energetic particles (ions) with atmospheric nuclei produce secondary particles; the most abundant are the pions (subatomic particle). The pions decay into muons and neutrinos. Muons reach the ground level and are detected by a variety of apparatus placed on the surface of the Earth and underground. The New-Tupi detector consists of four particle detectors. The four detectors are placed in pairs forming three telescopes. This configuration is able to measure the muon flux (coincidence rate between two detectors) from three directions: the vertical (zenith), west and east (with an inclination of 45°). The separation between the detectors (vertical and horizontal) is 2.83 m. Details of the New-Tupi detector can be found in [16].

The New-Tupi detector is within the South Atlantic Anomaly (SAA) region. This region is characterized by an anomalously weak geomagnetic field strength (less than 28,000 nT). It is in the SAA region where the inner Van Allen radiation belt makes its closest approach to the Earth's surface. This behavior of the magnetosphere is responsible for several processes, such as the high conductivity of the atmospheric layers due to the precipitation of energetic trapped particles from the inner Van Allen belt.

3. The St. Patrick's day storm

Two days before St Patrick's Day, on March 15, 2015, the sunspot 2297 produced a C9.1 class flare with onset at 01:15 UT and peaking at 02:13 UT. Indeed, a C-Flare does not draw attention because they are relatively common solar explosions. However, in this case, the blast was of long duration, beyond 2 hours and the active area was located at west of the solar disk (S22, W29), that is, in a geo-effective location. In addition, the blast triggered a partial halo coronal mass ejection (CME) toward the earth. The Solar Dynamics Observatory (SDO) [17] using the 94 angstroms channel has shown the explosion and it is reproduced in **Figure 2** (left panel), and the register of the CME by Cactus C2 Lasco is reproduced in **Figure 2** (right panel).

The detection of Type II and Type IV radio emission [18] was the indication of the presence of CME shocks, they are typically associated with strong coronal mass ejections and solar radiation storms. Indeed, the blast triggered a solar radiation storm, that is, solar energetic particles (SEP) began to flow around the planet at early hours, on March 15, 2015, on a gradual rise, reaching the condition of an almost S1 (minor) radiation storm in the NOAA scale on March 16. There was a small and narrow enhancement of the counting rate at the vertical New-Tupi telescope, in association with the SEP peak, as observed by ACE-SIS at L1 Lagrange point, as well as at GOES. **Figure 3** summarizes the situation.

According to WSA-ENLIL SOLAR WIND PREDICTION, the impact of the CME with the Earth's atmosphere would be only on March 18, 2015, triggering a geomagnetic storm of only G1 (minor) level of intensity. However, at 04:05 UTC on March 17, 2015, 15 hours earlier than expected, the ACE spacecraft detected the passage of a shock wave. In addition, the Bz component of the interplanetary magnetic field (IMF) was initially pointing north, but through

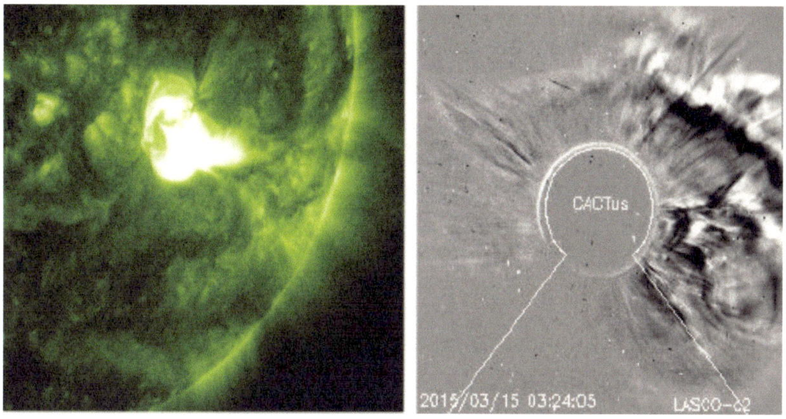

Figure 2. Left panel: Register of the long duration C9.1 peaking at 02:13 UT (March 15, 2015) around sunspot 2297. Credit: SDO/AIA. Right panel: Register of the partial CME associated to the C9.1 are. Credit: C2 Lasco.

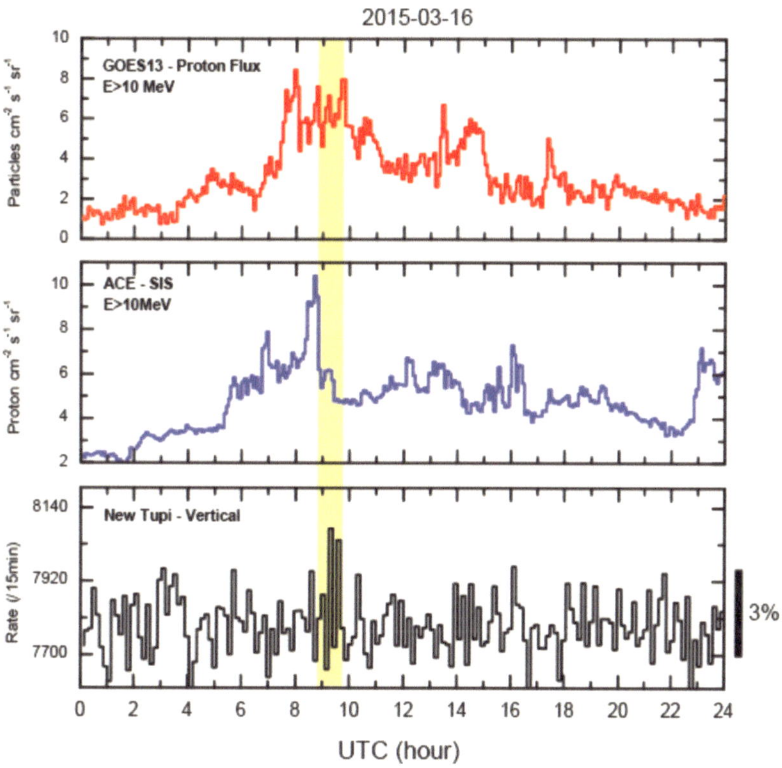

Figure 3. Time profiles observed in three different detectors on March 16, 2015. Top panel: the proton flux with energies above 10 MeV, according to the geostationary weather satellite. Central panel: the proton flux with energies above 10 MeV at Lagrange point L1, according to SIS instrument at ACE spacecraft. Bottom panel: the muon counting rate 15 minute binned, according to the vertical New-Tupi telescope.

a fast oscillation it finished pointing to the south with a maximum magnetic field (Bz) of −30 nT (−20 nT sustained). Then began a period of enhanced geomagnetic activity reaching G3 (strong) to G4 (severe) geomagnetic conditions and sustained for 12 hours, as shown in **Figure 4** (left top panel), where the time profiles of the interplanetary magnetic field Bt and Bz component are shown. Solar wind speeds as measured by ACE spacecraft increased to above 500 km/s, reaching up to a maximum wind speed of 680 km/sec, as shown in **Figure 4** (left central panel), where the effects of the CME on the solar wind at Lagrange point L1 is shown.

The arrival of the shock wave on Earth was 33 minutes after its passage through the Lagrange point L1, because there is the register at 04:35 UT of a geomagnetic sudden impulse measuring 54 nT by the USGS ground-based magnetometer in Boulder, Colorado. This signaled the moment the interplanetary shock wave (coronal mass ejection) swept past the Earth.

The bottom panel of **Figure 4** shows the register of the counting rate in two different instruments, New-Tupi detector, and Nain NM. Despite both being in different locations and consequently having different geomagnetic conditions, because the Nain NM is located at high latitude, close to the North Pole, their counting rates show a correlation. The FD intensity (changing in the counting rate) in New-Tupi was 2.5%, whereas in Nain was 3.8%. In addition, it is possible to see that the onset of FD observed at the muon counting rate in the New-Tupi telescope (and at the Nain NM) are in correlation with the onset of the geomagnetic storm.

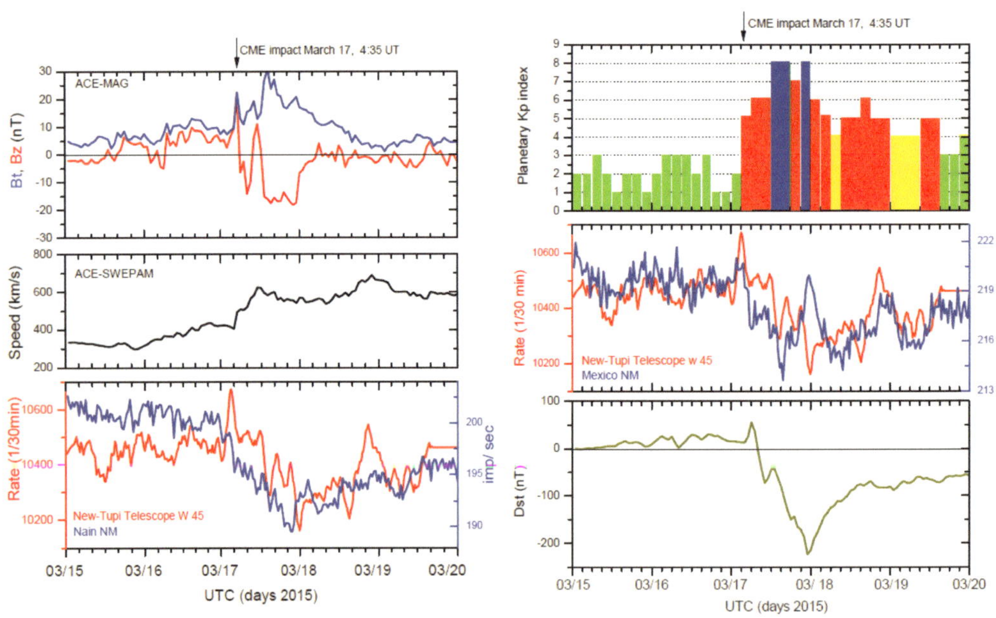

Figure 4. (Figures on the left) Top and central panels: the time profiles of the solar wind magnetic field components Bz and Bt and the solar wind speed, respectively, observed by the ACE spacecraft. Bottom panel: the 30 minutes muon counting rate in the New-Tupi mu telescope and Nain NM. (Figures on the right) Top panel: the time profiles of the geomagnetic planetary Kp index. Central panel: the 30 minutes muon counting rate in the New-Tupi muon telescope and Mexico NM. Bottom panel: the time profiles of the geomagnetic Dst index. In all cases, the time profiles are in the period from on March 15–19, 2015.

These signatures can be seen through the time profiles of the magnetic field, Bz, component (top panel) as well as with the solar wind speed increase (central panel).

The CME's impact on Earth triggered a G4 level (severe) (or Kp = 8) geomagnetic storm with onset about 14:00 UT on March 17, 2015, as is shown in the right top panel in **Figure 4**. The geomagnetic storm had a duration of about 18 hours, with a G3/G4 condition sustained for 12 hours. The right central panel in **Figure 4** shows also a good correlation between the time profiles of the Kp index and the counting rate in New-Tupi detector; in this case, we have included also, for comparison, the counting rate in Mexico NM. Finally, in the right bottom panel in **Figure 4** we have included the time profiles of the Dst index, based on the ring current around Earth produced by solar electrons and protons. They induce a magnetic field opposite Earth's magnetic field. During a geomagnetic storm, this induced magnetic field is higher than the Earth's geomagnetic field, and the resulting magnetic field is negative.

So far, this was the strongest G4 storm of Solar Cycle 24, known as the St. Patrick's Day storm.

4. The high-speed stream on October 7, 2015

Geomagnetic storms, at least the most intense, are generally the result of Coronal Mass Ejections intersecting with the Earth. However, coronal holes are the most common source of geomagnetic storms. Coronal holes occur when the Sun's magnetic field is open to interplanetary space; they are seen as a large dark region, and this characteristic is a consequence of being in colder regions. The open configuration of the magnetic field in coronal holes allows the particles to escape from the solar corona. Thus, they are sources of high-speed solar wind streams. Thus, when the coronal hole is geo-effective, that is, located near the Sun's central region, its associated HSS can reach the Earth's magnetosphere; under this condition there is an efficient exchange of energy from the high-speed solar wind to the magnetosphere, producing a disturbance on it.

An equatorial coronal hole high speed stream (HSS), with positive polarity flowed on 6 October 2015, could be seen in the SDO images, such as shown in **Figure 5**.

The HSS plasma flows following the spirals of the interplanetary magnetic field lines, forming a stream interaction region (SIR) at the compressed boundary between the fast and slow wind. **Figure 6** shows a schematic of this configuration, where an HSS with the southward interplanetary magnetic field sets up the interplanetary electric field $E = -V_{IMF} \times B$. The Parker spiral has a minor influence on the HSS than on the slower solar wind speed.

We would like to point out that on October 7, 2015 two coincident phenomena occurred; both contributed to the enhancement of the geomagnetic storm on October 7, 2015.

The first was the crossing of Earth for the Heliospheric Current Sheet (HCS), as represented in the top panel of **Figure 7**; it gives a snapshot of the WSA-Enlil model, run around on October 7, 2015 [16, 19, 20]. The model shows the solar wind plasma density in the ecliptic plane. Green to red colors show very high plasma density and blue color represents low density. A HCS is a

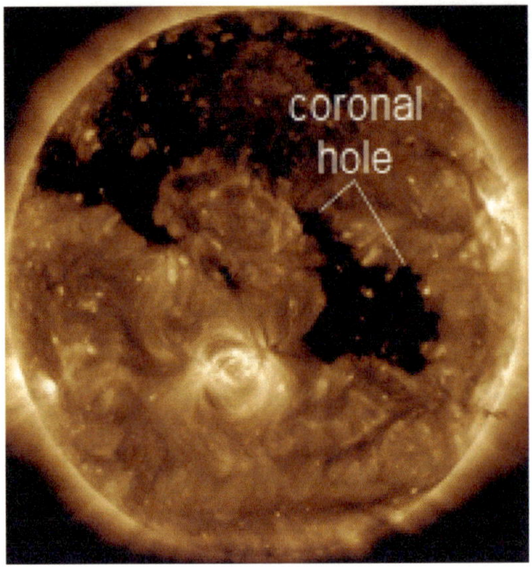

Figure 5. The sun equatorial coronal hole that gave origin to a high-speed stream (HSS), with positive polarity flowed on October 6, 2015. Credit: C2 SDO/AIA.

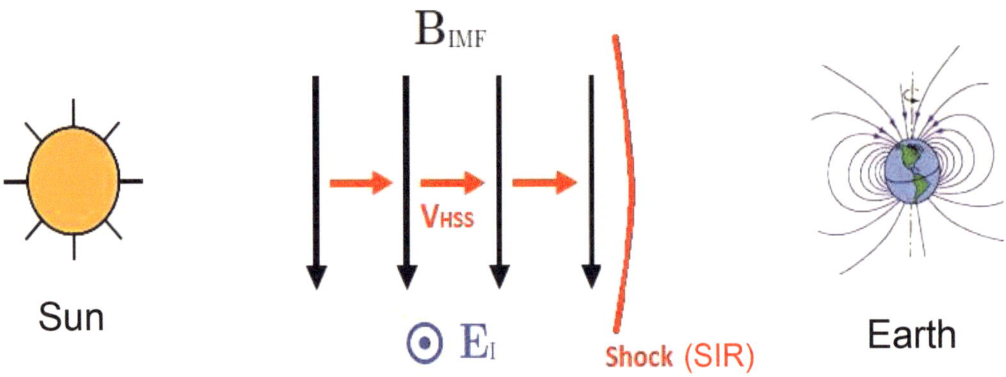

Figure 6. Schematic image of the high-speed stream approaching the Earth magnetosphere. The arrows pointing downwards represent the interplanetary magnetic field. In this configuration, an electric field E (open circle with dot) points out from the plane. The rightward arrows represent the solar wind speed. The arc line indicates a stream interaction region (SIR) forward shock.

transition zone that separates regions of opposing interplanetary magnetic field polarity [21]. An increase in the density of plasma concentration and of the interplanetary magnetic field (IMF) intensity is associated with the HCS compression region. This means that the HCS plays a magnetic focusing role to charged particles; they propagate more efficiently following the HCS sectors. In some cases, the crossing of Earth by HCS also can propitious a small enhancement in galactic cosmic ray (GCR) flux observed at ground level, as shown by [22].The Earth is marked as the green circle, and the Sun by the yellow circle. Following this figure, we can see

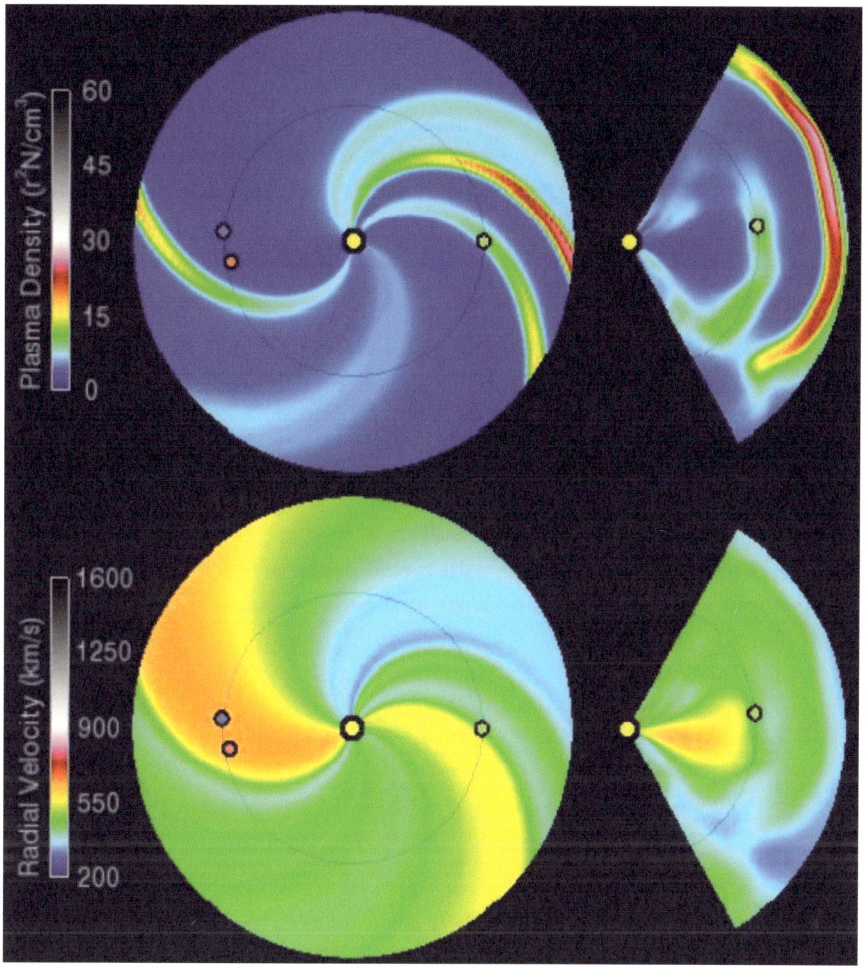

Figure 7. Top panel: Solar wind particle density and bottom panel: the solar wind speed gradients, according to the WSA-Enlil model. The central circle represents the Sun and the circle at the right represents the Earth on October 7, 2015.

that on 7 October the Earth was crossing a narrow region of high solar plasma density, that is, crossing the HCS.

The second was that at early hours of October 7, 2015, the Earth entered a region of a high-speed stream from a positive polarity coronal hole, as provided by the WSA-Enlil Solar Wind Prediction model and shown in the bottom panel in **Figure 7**. The long large-scale plasma structure, generated by the interaction of a stable fast solar wind stream, with the surrounding slow solar wind is called as corotating interaction region (CIR) [23]. The generation of forward and reverse magnetic shocks in the boundary regions of the CIR can accelerate ions from MeV to GeV energies. In addition, the magnetic perturbation of the magnetosphere due to the impact of the HSS can act as shadow effect to the galactic cosmic rays, producing an FD at ground-level detectors. Thus, the HSS impact on the magnetosphere on 7 October had the influence to start a period of an active geomagnetic storm.

The first signal of the HSS was at early hours, about 02:00 UTC at Lagrange Point L1, the interplanetary magnetic field had an enhancement of around twice its previous value, as shown in the ACE-MAGNETIC and reproduced in **Figure 8** (top left panel). Simultaneously, the Bz component had a southward deviation. In addition, the solar wind speed had a fast rise from 400 km/s to above 700 km/s and reproduced in **Figure 8** (central left panel). These were the signatures of an interplanetary shock wave due to the passage of HSS by the Lagrange Point L1.

In almost temporal coincidence with the fast rise of the solar wind speed at ~12:00 UT, there was a fast falling in the counting rate (FD) of the New-Tupi telescope that pointed to the west, with a confidence about 2.5%; the result is in **Figure 8** (bottom left panel). As predicted, the effect at Earth was a geomagnetic storm with onset at early hours on 7 October, reaching the condition of G3(strong) or Kp = 7 as shown in **Figure 8** (top right panel). It was a geomagnetic storm of relatively long duration, covering October 7 and 8 with a G2/G1 (Kp = 6/Kp = 5) conditions, sustained for more of 21 hours.

In addition, from **Figure 8** we can see that the FD observed in the muon counting rate of New-Tupi telescope (central right panel) was with a confidence of 2.5%. It is correlated with the geomagnetic storm, whose signature at Earth can be seen through the time profiles of the planetary Kp index (top right panel). As well as in the Dst index time profiles (bottom right

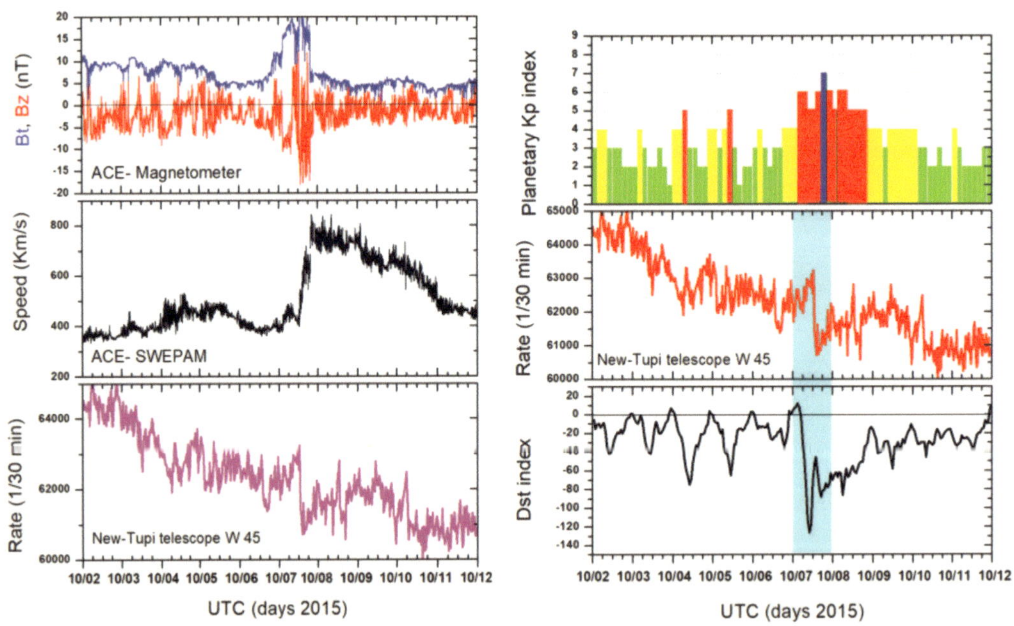

Figure 8. (Figures on the left) Top and central panels: the time profiles of the solar wind magnetic field components Bz and Bt and the solar wind speed, respectively, observed by the ACE spacecraft. Bottom panel: the 30 minutes muon counting rate in the New-Tupi muon telescope. (Figures on the right) Top panel: the time profiles of the geomagnetic planetary Kp index. Central panel: the 30 minutes muon counting rate in the New-Tupi muon telescope. Bottom panel the time profiles of the geomagnetic Dst index. In all cases, the time profiles are in the period from on October 2–12, 2015.

panel). Notice that the fast fall at New-Tupi coincides with the minimum Dst value, reaching a value of about −120 nT.

We would like to point out, that a similar behavior, such as the one observed at New-Tupi also was observed in the counting rate of the South Pole NM. However, in most NMs, the fall in the counting rate does not present a clear FD. In most cases it is only a fall smaller than 1.5% and with a recovery time less than 6 hours.

5. Impact of solar activity in the terrestrial climate

5.1. Incidence of electromagnetic radiation

The principal source of energy on Earth is the Sun. The radiated energy by the Sun covers all electromagnetic spectrum. However, the radiation on the range of the visible spectrum (wavelengths around 400–800 nm) has the most contribution. The Sun's visible radiation at Earth (distance of 1-AU) is nearly constant, and the main variation of around 0.1% follows the 11-year solar cycle progression. The total wavelength-integrated energy from sunlight at Earth is known as the Solar Constant (SC) or Total Solar Irradiance (TSI) whose value vary from 1365.5 Watts/m^2 at solar minimum to 1366.5 Watts/m^2 at solar maximum.

There is also a minor contribution of the energy irradiated by the Sun in other wavelengths of the electromagnetic spectrum, such as the Ultraviolet (UV) (wavelengths around 120–400 nm), the X-rays and gamma radiations. The radiations with wavelengths smaller than the wavelength of visible radiation are absorbed in the upper atmosphere, for instance, a large fraction of UV radiation is absorbed by ozone in the stratosphere. The gamma and X-ray radiations have large variability of up to 300% over very short timescales (e.g., several minutes). This large variability is associated with solar transient events such as flares. These radiations in short wavelength have minimal impact on the Earth weather.

On the other hand, in wavelengths smaller than the visible spectrum, there is also a minor contribution, such as the Infrared (IR) (wavelengths 800–10,000 nm) with small changes of up to 1% over the solar cycle progression. Finally the solar radio waves, they are a transient contribution linked to the solar flares and they are called Radio Blackouts (RB). Depending on the solar flare intensity, RB can cover a wide area blackout of HF radio communication, loss of radio contact with duration of up to hours on the sunlit side of Earth and the degradation of Low-frequency navigation signals with a duration of up to hours.

5.2. Incidence of solar particles

The Earth geomagnetic field is a natural shielding to charged particles coming from space; low-energy particles are largely deflected when their incidence is near equatorial regions. However, the deflection is lower near Polar Regions or above roughly 55° of latitude.

In addition, the Sun is also the source of e Solar Energetic Particles (SEP) accelerated up to GeV energies during the occurrence of solar flares and CMEs. In space weather physics, these particles are the so-called radiation storms. If the solar active region is close to the central meridian (i.e., geo-effective region), the SEP can reach the Earth. The S1 level of a radiation storm indicates that the flux of protons at 1-AU with energies above 10 MeV equals or exceeds 10 pfu (1 pfu = 1 particle per cm^2 s sr). The scale is logarithmic, this means that 100 pfu indicate an S2 level, and so on.

Under certain conditions, SEP can penetrate into the magnetosphere and reach the upper and middle atmosphere. This behavior can have long lasting consequences, such as the changes in the concentration of some molecules as the NOx species [24]. However, the influence of these changes in the Earth weather is not yet clear.

5.3. Incidence of galactic cosmic rays

The interplanetary magnetic field (IMF) is modulated by the solar cycle; the maximum activity solar occurs every 11 years and it is when the Sun's magnetic field inverts its polarity and shuffles the IMF. This behavior modulates the lower energy component of the galactic cosmic rays (GCR), that is, the ~11-year variation in cosmic ray intensity observed at the Earth is anti-correlated with solar activity. The almost isotropic flux of GCR when they enter the helio-sphere is modified, around the solar maximum activity, their dispersion is higher, and this reduces the incidence of cosmic rays flux on the Earth. In addition, around the minimum solar activity, the Sun's magnetic field has a defined polarity and the cosmic rays dispersion is lower; consequently, the incidence of cosmic rays flux on the Earth is higher.

Thus, at solar minimum, the number of cosmic rays reaching Earth increases. These cosmic ray particles can be the source for creating nucleation sites in the atmosphere, that is, the seed of nucleation in cloud formation and create rain conditions [25]. In other words, the impact on Earth weather conditions can be modulated by the 11-year solar cycle. This topic needs further studies.

5.4. Injection of accelerated particles by an interplanetary shock

According to satellite observations, such as the NASA's IMAGE spacecraft and the joint NASA/European Space Agency Cluster satellites, immense cracks sometimes develop in Earth's magnetosphere and remain open for hours due to magnetic reconnection.

In the Earth, the magnetic reconnection is a phenomenon that occurs in several situations, in all cases, a high speed magnetized plasma, reach a region of the magnetosphere with an oppo sitely directed magnetic field, the components break and there is a cross-connect, forming a crack in the magnetosphere, transferring the energy stored in the magnetic fields to plasma motion and heat.

Reconnection can occur between Earth's magnetic field and the interplanetary magnetic field (IMF) carried by a coronal mass ejection or a high-speed stream, emanating from the sun. Reconnection occurs at the boundary of Earth's magnetosphere, the region of space that is dominated by Earth's magnetic field. However, despite some models predicting the properties of reconnection at this boundary, there are some issues, which are poorly understood, such as, where reconnection really occurs.

The model with a southward interplanetary magnetic field leads to a natural explanation of neutral points of the geomagnetic field [26]. A number of methods have been developed to locate magnetic separators, that is, the location where B = 0, such as the magnetic nulls in global magnetospheric simulations [27] that allow determining the separator for both northward and southward IMF orientations [28].

The first point of reconnection is in the day-side of Earth (left point in **Figure 9**) where a crack opens in the magnetic shield and remains open for hours, through which the electrically charged particles from space can flow. This includes particles from high-speed solar wind, particles accelerated in the front of the shock wave and up to galactic cosmic rays of low energy.

There is also reconnection in the magneto-tail results in a cross-tail electric field, in the night-side of Earth (right point in **Figure 9**). The electric field generated by this reconnection process directs the plasma injected by the crack due to the reconnection in the day-side, again to the Earthward.

At about three times of the Earth radii, the ions and electrons begin drifting in different directions around the Earth; in the geomagnetic equatorial plane, the positive ions drift westward, and the electrons drift eastward. This charge separation creates the ring current J_{DST}. The ring current around Earth produces a magnetic field B_{DST} that is directly opposite the Earth's magnetic field and generates the so-called Dst index. A negative Dst value means that Earth's magnetic field is weakened. This is particularly the case during solar magnetic storms.

Recent results [16] show some evidences of a particle injection into the Earth atmosphere, driven by the HSS on August 28, 2015, as shown in **Figure 10**. In the top panel we show the time profiles of the New-Tupi muon counting rate, the central and bottom panels show the time profiles of the solar wind speed and particle density in the ACE spacecraft, respectively. The impact of HSS (shock arrival) occurred when the magnetosphere was already disturbed by the impact of a previous CME on October 25, 2015, triggering a geomagnetic storm and a FD that was already underway.

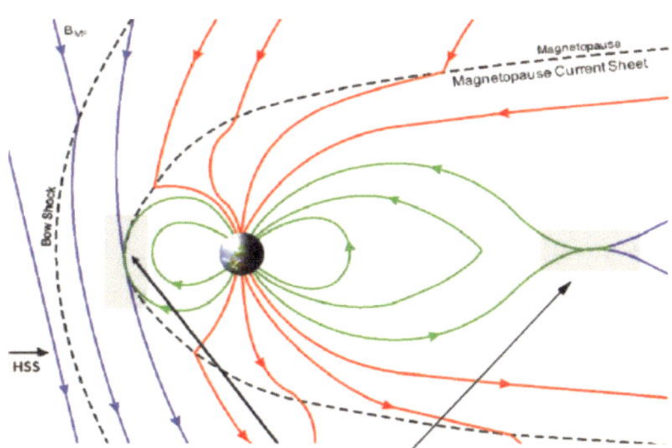

Figure 9. Schematic view of the magnetic reconnection in the Earth's magnetosphere. The black arrows indicate where magnetic reconnection occurs.

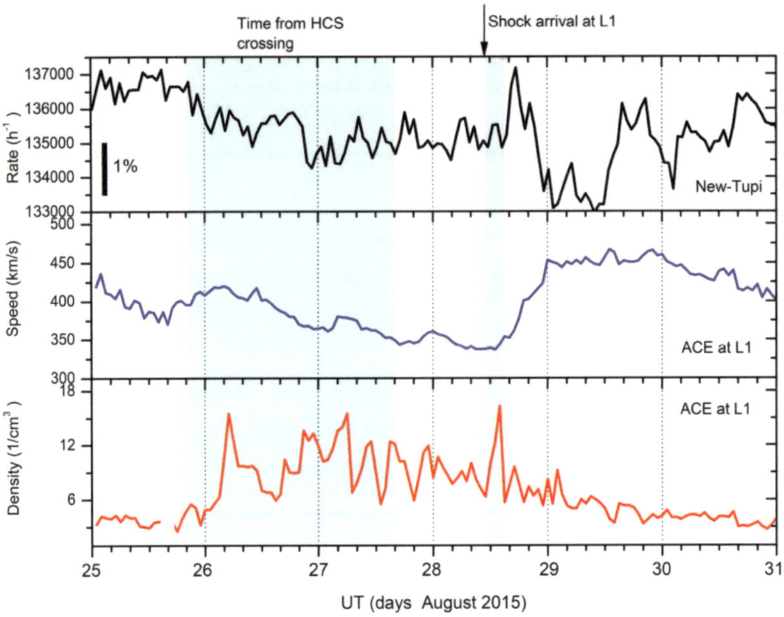

Figure 10. Comparison between the muon counting rate in the New-Tupi telescope (top panel) and the solar wind parameters (density and speed) observed by the ACE satellite (center and bottom panels) in the period from August 25 until August 31. The shaded area indicates the HCS crossing time and the vertical arrow show the arrival of a shock associated with a high-speed stream on August 28.

The energy of the injected particles reached up to GeV values because from their interactions in the upper atmosphere, muons were produced and a fraction of them reached the ground level. In most cases, these injected particles are directed to the Polar Regions. However, the particles from the high-energy tail of the spectrum can reach mid-latitudes. The injected energy in the upper atmosphere during the HSS impact had a duration of ~350 minutes and a value above 1.4×10^{-4} erg per cm^2. This behavior could have consequences for the atmospheric chemistry. For instance, the creation of NOx species may be enhanced, and this can lead to increased ozone depletion [24]. This topic requires further study.

6. Conclusion(s)

Space magnetic conditions in the interplanetary medium and consequently within the Earth's magnetic environment are constantly influenced by the transient disturbances in the interplanetary medium propagating from the Sun, such as CMEs and HSSs. We have presented here two examples of these events producing magnetic disturbances known as geomagnetic storms.

The first is one of the highest geomagnetic storms in the current solar cycle 24, on March 17, 2015 (St. Patrick's Day) and it reached the condition of G4 (Kp = 8) in the NOAA storm scale. The origin was a CME impact, consequently. The second was the impact of an HSS on October

7, 2015, whose origin was a positive coronal hole. In both cases, the passage of these two events by L1 had an interplanetary shock signature, with a fast rising of the solar wind. The effect at ground level was an FD; however, it is possible to see (especially in the first event) some particle enhancements. The probable origin of these particles' enhancement is the anisotropies in the perturbed Earth magnetic field, the magnetic reconnection between the magnetic field of the interplanetary shock and the geomagnetic field, opening cracks in the Earth magnetic shield (please see Section 5).

In order to see better the particle injection in the upper atmosphere, we show a prolonged energetic (in the GeV energy scale) particle injection within the Earth atmosphere system, driven by an HSS on August 28, 2015.

This mechanism could be correlated with the renovation of the atmospheric chemistry. In addition, the formation of NOx species can be enhanced and can be correlated with the ozone depletion and abundance [24]. This mechanism correlates with the atmospheric chemistry requires further study.

Acknowledgements

The support from National Council Research (CNPq) and Fundação de Amparo a Pesquisa do Estado do Rio de Janeiro (FAPERJ) both in Brazil is gratefully acknowledged. We express our thanks to the ACE/MAG instrument team, the ACE Science Center, the NASA GOES team and the NOAA Space Weather Prediction Center (www.swpc.noaa.gov), the Geomagnetism at Kyoto University, Japan (wdc.kugi.kyoto-u.ac.jp) and the NMDB database (www.nmdb.eu), founded under the European Union's FP7 program (contract no. 213007), for providing real time data. We also express our gratitude to V. Kopenkin and A. Nepomuceno for valuable discussions and their help.

Author details

Carlos E. Navia*, Marcel N. de Oliveira and Carlos R. A. Augusto

*Address all correspondence to: tupi.carlos24@gmail.com

Instituto de Física, Universidade Federal Fluminense, Rio de Janeiro, Brazil

References

[1] Rezaei R, Schmidt W, Beck C. Comparison of sunspot properties in cycles 23 and 24. In: 40th COSPAR Scientific Assembly. 2014;**40**:E2.4-15-14. ADS: 2014cosp…40E2740R

[2] Basu S. The peculiar solar cycle 24–where do we stand? Journal of Physics: Conference Series. IOP Publishing. 2013;**440**:012001. DOI: 10.1088/1742-6596/440/1/012001

[3] Abdo A, Ackermann M, Ajello M, Baldini L, Ballet J, Barbiellini G, et al. Fermi large area telescope observations of two gamma-ray emission components from the quiescent sun. The Astrophysical Journal. 2011;**734**(2):116. DOI: 10.1088/0004-637X/734/2/116

[4] Hathaway DH. The solar cycle. Living Reviews in Solar Physics. 2015;**12**:4. DOI: 10.1007/lrsp-2015-4

[5] Gonzalez WD, Gonzalez AC, Tsurutani BT. Dual-peak solar cycle distribution of intense geomagnetic storms. Planetary and Space Science. 1990;**38**(2):181. DOI: 10.1016/0032-0633(90)90082-2

[6] Vaquero JM, Kovaltsov GA, Usoskin IG, Carrasco VMS, Gallego MC. Level and length of cyclic solar activity during the maunder minimum as deduced from the active-day statistics. A&A. 2015;**577**:A71. DOI: 10.1051/0004-6361/201525962

[7] Gonzalez WD, Joselyn JA, Kamide Y, Kroehl HW, Rostoker G, Tsurutani BT, Vasyliunas VM. What is a geomagnetic storm? Journal of Geophysical Research. 1994;**99**:5771. DOI: 10.1029/93JA02867

[8] Augusto CRA, Kopenkin V, Navia CE, Tsui KH, Shigueoka H, Fauth AC, et al. Variations of the muon flux at sea level associated with interplanetary ICMEs and corotating interaction regions. The Astrophysical Journal. 2012;**759**:143. DOI: 10.1088/0004-637X/759/2/143

[9] NOAA / NWS Space Weather Prediction Center. NOAA Space Weather Scales [Internet]. 2018. Available from: http://www.swpc.noaa.gov/sites/default/files/images/NOAAscales.pdf [Accessed: 26-January-2018]

[10] Forbush SE. On the effects in cosmic-ray intensity observed during the recent magnetic storm. Physics Review. 1937;**51**(12):1108. DOI: 10.1103/PhysRev.51.1108.3

[11] Cane HV. Coronal mass ejections and Forbush decreases. In: Cosmic Rays and Earth. Space Science Reviews. 2000;**93**:55. DOI: 10.1023/A:1026532125747

[12] Lockwood JA, Webber WR, Debrunner H. Forbush decreases and interplanetary magnetic field disturbances: Association with magnetic clouds. Journal of Geophysical Research. 1991;**96**(A7):11587. DOI: 10.1029/91JA01012

[13] Belov A, Abunin A, Abunina M, Eroshenko E, Oleneva V, Yanke V, et al. Coronal mass ejections and non recurrent Forbush decreases. Solar Physics. 2014;**289**:3949. DOI: 10.1007/s11207-014-0534-6

[14] Kavanagh A, Denton M. High-speed solar-wind streams and geospace interactions. Astronomy & Geophysics. 2007;**48**:6.24. DOI: 10.1111/j.1468-4004.2007.48624.x

[15] Tsurutani BT, Gonzalez WD, Gonzalez AL, Guarnieri FL, Gopalswamy N, Grande M, et al. Corotating solar wind streams and recurrent geomagnetic activity: A review. Journal of Geophysical Research. 2006;**111**(A7). DOI: 10.1029/2005JA011273

[16] Augusto CRA, Navia CE, de Oliveira MN, Nepomuceno AA, Kopenkin V, Sinzi T. Muon excess at sea level during the progress of a geomagnetic storm and high-speed stream

impact near the time of Earth's Heliospheric sheet crossing. Solar Physics 2017;**292**:107. DOI: 10.1007/s11207-017-1116-1

[17] Lemen JR, Title AM, Akin DJ, et al. The atmospheric imaging assembly (AIA) on the solar dynamics observatory (SDO). Solar Physics. 2012;**275**:17. DOI: 0.1007%2Fs11207-011-9776-8

[18] Stewart RT, Sheridan KV. Evidence of type II and type IV solar radio emission from a common flare-induced shock wave. Solar Physics. 1970;**12**:229. DOI: 10.1007/BF00227120

[19] Odstrcil D, Smith Z, Dryer M. Distortion of the heliospheric plasma sheet by interplanetary shocks. Geophysical Research Letters. 1996;**23**:2521. DOI: 10.1029/96GL00159

[20] Odstrcil D. Modeling 3-D solar wind structure. Advances in Space Research. 2003;**32**:497. DOI: 10.1016/S0273-1177(03)00332-6

[21] Wilcox JM, Ness NF. Quasi-stationary corotating structure in the interplanetary medium. Journal of Geophysical Research. 1965;**70**(23):5793. DOI: 10.1029/JZ070i023p05793

[22] Thomas SR, Owens MJ, Lockwood M, Scott CJ. Galactic cosmic ray modulation near the heliospheric current sheet. Solar Physics. 2014;**289**(7):2653. DOI: 10.1007/s11207-014-0493-y

[23] Pizzo V. A three-dimensional model of corotating streams in the solar wind. Journal of Geophysical Research. 1978;**83**(A12):5563. DOI: 10.1029/JA083iA12p05563

[24] Seppala A, Verronen PT, Clilverd MA, Randall CE, Tamminen J, Soeva V, Kyrolo E. Arctic and Antarctic polar winter NOx and energetic particle precipitation in 2002–2006. Geophysical Research Letters. 2007;**34**:L12810. DOI: 10.1029/2007GL029733

[25] Kirkby J, Curtius J, Almeida J, Dunne E, Duplissy J, Ehrhart S, et al. Role of sulphuric acid, ammonia and galactic cosmic rays in atmospheric aerosol nucleation. Nature. 2011;**476**(7361): 429. DOI: 10.1038/nature10343

[26] Dungey JW. Interplanetary magnetic field and the auroral zones. Physical Review Letters. 1961;**6**(2):47. DOI: 10.1103/PhysRevLett.6.47

[27] Dorelli JC, Bhattacharjee A, Raeder J. Separator reconnection at Earth's day-side magnetopause under generic northward interplanetary magnetic field conditions. Journal of Geophysical Research. 2007;**112**:A02202. DOI: 10.1029/2006JA011877

[28] Hu YQ, Peng Z, Wang C, Kan JR. Magnetic merging line and reconnection voltage versus IMF clock angle: Results from global MHD simulations. Journal of Geophysical Research. 2009;**114**:08220. DOI: 10.1029/2009JA014118

Chapter 4

Rainfall Distribution in Landfalling Tropical Cyclones

Zifeng Yu and Yuqing Wang

Additional information is available at the end of the chapter

http://dx.doi.org/10.5772/intechopen.75910

Abstract

Accurate prediction of rainfall distribution in landfalling tropical cyclones (LTCs) is very important to disaster prevention but quite challenging to operational forecasters. This chapter will describe the rainfall distribution in LTCs, including both axisymmetric and asymmetric distributions and their major controlling parameters, such as environmental vertical wind shear, TC intensity and motion, and coastline. In addition to the composite results from many LTC cases, several case studies are also given to illustrate the predominant factors that are key to the asymmetric rainfall distribution in LTCs. Future directions in this area and potential ways to improve the operational forecasts of rainfall distribution in LTCs are also discussed briefly.

Keywords: tropical cyclones, landfalling, rainfall distribution, vertical wind shear, typhoon, precipitation, intensity

1. Introduction

Landfalling tropical cyclones (LTCs) often bring very heavy rainfall to the affected regions so that many extreme rainfall events are related to LTCs around the world. Heavy rainfall associated with LTCs is one of the most devastating natural disasters in the coastal regions in the world, which could bring unexpected flash floods and landslides to inflict huge losses in properties and human lives. Examples of some record-breaking heavy rainfall events induced by LTCs include 24-h rainfall of 1062 mm at Linzhuang in Henan Province of Mainland China caused by landfalling Typhoon Nina (7503) in 1975 and 24-h rainfall of about 1300 mm in southern Taiwan caused by Typhoon Morakot in 2009. Therefore, the rainfall distribution in a LTC is of particular importance to meteorologists and disaster preventions.

However, though tropical cyclone (TC) track forecasts have achieved great improvements in the past decades [1, 2], improvements in TC rainfall forecasts have been largely behind. Even when the TC track is well predicted, the rainfall distribution and intensity are often poorly predicted, especially for rainfall distribution and storm structure. Therefore, understanding the characteristics and the factors that control rainfall distribution in LTCs is very important for improving real-time operational rainfall forecasts.

Distribution of rainfall in LTCs may be affected by many processes and factors, such as TC motion, intensity, landfalling location and time, large-scale environmental conditions, terrains in the coastal region, and so on. Previous studies have shown that rainfall in a TC can be largely modulated by the environmental vertical wind shear (VWS) [3–12], the planetary vorticity gradient [13–16], the storm motion [16, 17], the friction-induced asymmetric boundary layer convergence in a moving TC [18], the asymmetric environmental moisture distribution in [19], and also the convectively coupled vortex-Rossby waves in the eyewall [20–22].

To investigate the rainfall distribution in LTCs and the associated physical process, the rainfall distribution in a LTC is often decomposed into a wavenumber (WN)-0 (or an azimuthal mean) component and a series of higher-WN components in [23–26]. Recently, with the improvements of satellite-retrieved rain data application, some observational studies have focused on the asymmetric TC rainfall distribution and related different physical processes. Rogers et al. [27] suggested that the combination of VWS and storm motion could explain well the rainfall asymmetry in Hurricane Bonnie. Lonfat et al. [23] analyzed the global rainfall distribution in TCs based on the Tropical Rainfall Measuring Mission (TRMM) rain estimates and studied the relationships between the TC rainfall distribution and the TC motion over oceans in the world. It has been shown that the storm motion alone, however, could not fully explain the basin-to-basin variability in the rainfall asymmetry, suggesting that other mechanisms must contribute to the asymmetric rainfall distribution, such as the environmental VWS, which has been investigated extensively (e.g., [14, 24, 28]).

Chen et al. [24] investigated the effects of both VWS and storm motion on TC rainfall asymmetries based on composite analyses using the TRMM rain data. The results showed that the TC rainfall asymmetry generally depends on the juxtaposition and relative magnitude of the environmental VWS and the TC motion over the open ocean. Subsequent composite analyses of the TC rainfall asymmetry related to environmental VWS by using the satellite-based rainfall estimates have further confirmed the relationship between the spatial distribution of TC rainfall and the environmental VWS [29–31]. Namely, the maximum rainfall predominantly occurs in the downshear-left quadrant in a TC that is embedded in an environmental VWS (facing down the shear vector), while the effect of TC translation becomes an important factor for a TC embedded in the relatively low VWS environment. Reasor et al. [32] reconfirmed that the environmental VWS is the primary factor affecting the eyewall convective asymmetry in TCs based on Doppler radar data composite analysis.

Most of previous studies have mainly focused on the rainfall distributions in TCs before landfall or over the open oceans. In addition to the effects of environmental VWS, TC intensity and motion, the rainfall asymmetries in a LTC may also be affected by non-uniform surface characteristics, such as land-sea contrast, coastline, topography, and mesoscale convective activities that resulted from interactions with other synoptic weather systems [33–47].

Some numerical studies have found that the asymmetric structure of a LTC and the associated asymmetric rainfall distribution could be forced by land-sea contrast [20, 48–52]. Based on idealized simulations for LTCs on an *f*-plane, Li et al. [43, 44] showed that rainfall asymmetry became larger after landfall with larger VWS, and land surface roughness would initially affect rainfall distribution in the outer region. Xu et al. [45] indicated that the asymmetric rainfall percentage toward the right quadrant relative to the coastline would have an obvious increase in LTCs, which might be related to the land-sea roughness gradient.

In an early observational study, based on the airborne radar reflectivity data from systems onboard three National Oceanic and Atmospheric Administration (NOAA) aircrafts during 6 days for Hurricane Allen (1980), Marks [3] showed that changes in both the storm intensity and the eyewall radius had little impact on the total rainfall of the storm. However, the axisymmetric rainfall distribution in TCs has received further attention in recent years, with the focus on the relationship between TC intensity and axisymmetric rainfall distribution. Lonfat et al. [23] explored the axisymmetric rainfall distribution in TCs with different intensities during 1998–2000 over global oceans using the TRMM retrieved rain data. They found that the peak radial axisymmetric rain rate increased with increasing TC intensity. More recent studies have confirmed that TC intensity and intensity change are closely related with rainfall and rainfall change over the open ocean [53–57]. However, the relationship of axisymmetric rainfall distribution and TC intensity in LTCs is reported to be more complicated in Yu et al. [26].

This chapter aims to document the evolution of rainfall distribution in LTCs, taking those that made landfall over China as examples, and to systematically examine the effects of various factors, including TC intensity and motion, environmental VWS, and land-sea contrast, on both the axisymmetric and asymmetric rainfall distributions from 24 h prior, to landfall, and to 24 h after landfall. Section 2 introduces the data used and the analysis methods. Sections 3 and 4 present, respectively, the axisymmetric and asymmetric rainfall distributions in LTCs and their relationship with the different factors. Section 5 shows several case studies to demonstrate the complexity of rainfall distribution in LTC. Some remaining issues are discussed in Section 6. Section 7 gives the concluding remarks.

2. Data and methodology

2.1. TRMM 3B42 data

With the development of satellite observing systems, the satellite-retrieved rainfall estimates are more widely applied to both weather and climate research, in particular for the study of rainfall in TCs either over the open oceans or near landfall. Though rain gauge data are routinely available, they are not dense enough or even spare in many important regions, such as over terrains and in the coastal areas over land. Weather radars could give good temporal and spatial resolutions, but they are limited by their area coverage due to unevenly distributed radar sites and inter-radar calibration that could lower the capability and accuracy in rainfall estimation. Therefore, the satellite rainfall estimates are good choices to study rainfall characteristics, especially over the open oceans and coastal and mountainous areas where surface observations are limited.

Many studies have evaluated the performances of TRMM satellite rainfall retrievals in capturing the rainfall features in LTCs [58–62]. These studies have shown that the TRMM 3B42 rain data can provide quite reasonable rainfall distributions in LTCs compared with surface data like gauge and radar observations, although they are generally more accurate over ocean than over land. The rainfall distribution evolution in LTCs during 24 h prior to landfall (t = −24) to 24 h after landfall (t = 24) over China will be discussed in Sections 3 and 4 based on the TRMM 3B42 rain estimates.

2.2. Reanalysis data

The horizontal winds obtained from NCEP/NCAR (National Centers for Environmental Predictions/National Center for Atmospheric Research) global reanalysis data [63] are used to calculate the large-scale environmental VWS, which was defined as the 200-hPa and 850-hPa wind vector differences averaged within a 500-km radius from each TC center at every 6-h interval as in [25, 26].

2.3. Best-track TC data

In total, 133 LTCs over China are included in the results discussed in the following sections (**Table 1**), with the lowest intensity of tropical storm (CAT2) from 24 h before landfall (t = −24), to landfall time (t = 0), and to 24 h after landfall (t = 24). Among them, 24, 39, 32, 29, and 9 LTCs, respectively, made a landfall in Hainan (HN), Guangdong (GD), Taiwan (TW), Fujian (FJ), and Zhejiang (ZJ) provinces in China (**Figure 1a–e**). The coastline directions of the five regions are also shown in **Figure 1f**, namely, 45, 16, 76, 53, and 45° from the east, respectively, in HN, GD, TW, FJ, and ZJ.

The best-track TC data are those products from the Shanghai Typhoon Institute (STI) of China Meteorological Administration (CMA), including intensities (the maximum near-surface 2-min

Time	**TS (CAT = 2, 17.2–24.4 m s^{-1})**	**STS (CAT = 3, 24.5–32.6 m s^{-1})**	**TY (CAT = 4, 32.7–41.4 m s^{-1})**	**STY (CAT = 5, 41.5–50.9 m s^{-1})**	**SuperTY (CAT = 6, ≥51 m s^{-1})**
−24 h (t = −24)	30 (26.09%)	14 (12.17%)	39 (33.91%)	19 (16.52%)	13 (11.31%)
−18 h (t = −18)	27 (22.69%)	22 (18.48%)	39 (32.77%)	20 (16.81%)	11 (9.25%)
−12 h (t = −12)	28 (23.33%)	20 (16.67%)	40 (33.33%)	22 (18.34%)	10 (8.33%)
−6 h (t = −6)	23 (19.17%)	24 (20.00%)	41 (34.17%)	26 (21.66%)	6 (5.00%)
0 h (t = 0)	28 (24.13%)	25 (21.55%)	37 (31.90%)	22 (18.97%)	4 (3.45%)
6 h (t = +6)	42 (39.62%)	32 (30.19%)	28 (26.42%)	3 (2.83%)	1 (0.94%)
12 h (t = +12)	46 (51.11%)	25 (27.78%)	18 (20.00%)	0	1 (1.11%)
18 h (t = +18)	42 (59.15%)	17 (23.94%)	11 (15.49%)	1(1.41%)	0
24 h (t = +24)	40 (72.73%)	8 (14.54%)	7 (12.73%)	0	0

Note: The intensity is defined as the maximum sustained 10-m wind speed.

Table 1. Samples and frequencies of TCs with intensity of five categories (CAT = 2–6) from 24 h prior to landfall (t = −24), at the time of landfall (t = 0), to 24 h after landfall (t = 24) at every 6 h.

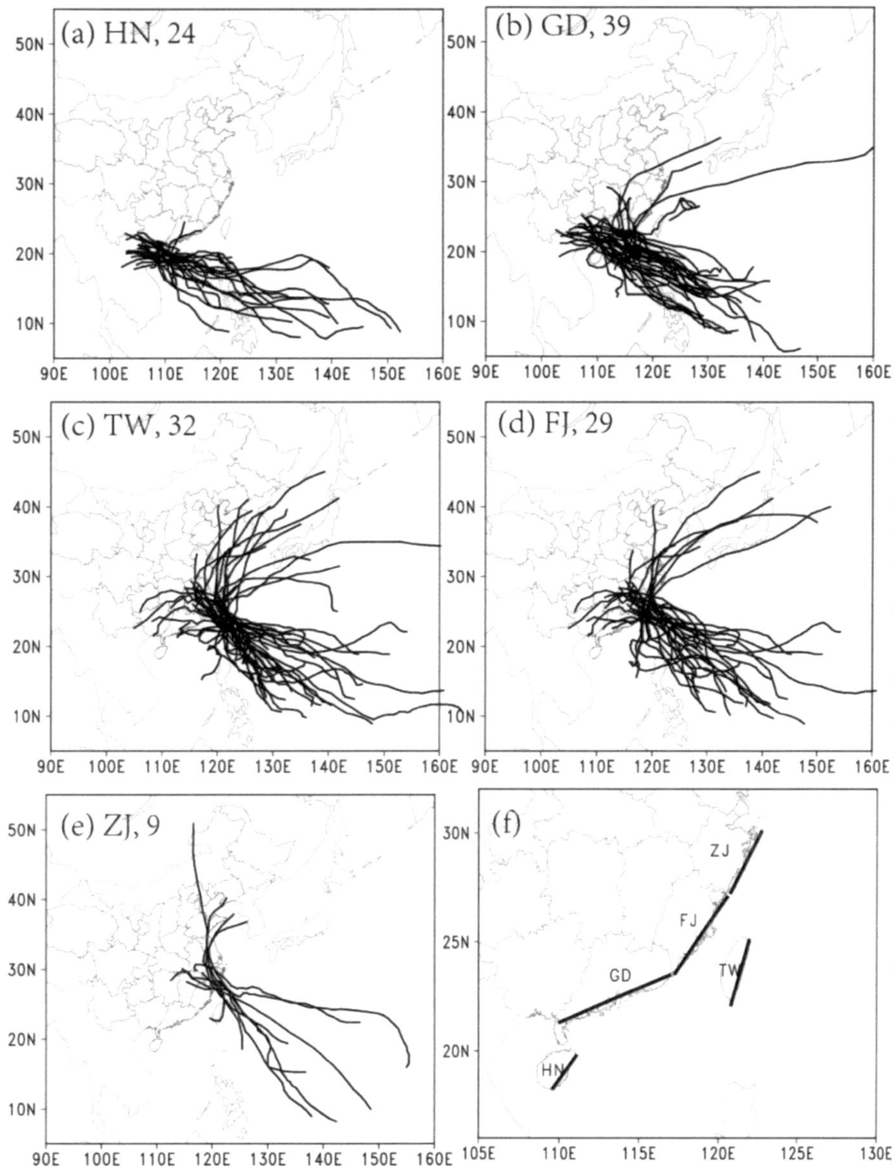

Figure 1. Tracks of TCs that made landfall in different regions (provinces) over China during 2001–2015: (a) Hainan (HN), (b) Guangdong (GD), (c) Taiwan (TW), (d) Fujian (FJ), and (e) Zhejiang (ZJ). Numbers of TCs making landfall in individual regions are illustrated in each panel. The names and locations of the five regions and their coastlines are shown in (f).

mean wind speed and the minimum sea-level pressure) and positions (longitude and latitude) of TCs at every 6-h interval during 2001–2015. **Table 1** shows the related 6-h intensity information including category, sample, and frequency of LTCs over China from t = −24 h to t = 24 h. The motion direction and speed of each TC at every TRMM observational time are calculated through the two nearest best-track records at 6-h interval. The intensity of a TC at landfall time is obtained by interpolation using the maximum near-surface wind speed in the best-track data at 6-h interval.

Table 2 lists the intensity change information (including the calculation methods and samples). The intensity changes of LTCs are categorized into five groups: rapidly decaying (RD), slowly decaying (SD), unchanged, slowly intensifying (SI), and rapidly intensifying (RI), according to the rate of TC intensity change. The RI threshold of 15 m s^{-1} is defined as the 24-h maximum wind speed change, while the RD threshold is defined as −15 m s^{-1} in the 24-h intensity change [64]. As a result, the TCs of RI, SI, unchanged, SD, and RD have the 24-h intensity changes of ≥15, (0, 15), 0, (−15, 0), and ≤−15 m s^{-1}, respectively. Note that the unchanged category is exactly defined as 0 m s^{-1}, and it occurs rather frequently, particularly within 24 h before landfall.

2.4. Rainfall analysis method

Fourier decomposition, which can help to compute the WN-0 and higher WNs (e.g., 1–4) rainfall components in a TC, is often used in the studies of rainfall distributions in TCs [23, 24]. The axisymmetric rainfall component is the azimuthally mean rain rate (WN-0) as a function of the radial distance in 10-km-wide annuli from the TC center. The asymmetric rainfall components were analyzed by binning rainfall rate data in 10-km-wide annuli from each TC center to 500-km radius. First, the WN-1 Fourier coefficients were computed by using all rain rate estimates as in Boyd [65]:

$$a_1 = \sum_i [R_i \cos(\theta_i)], b_1 = \sum_i [R_i \sin(\theta_i)], \tag{1}$$

Time	RI (ΔV ≥ 15 m s^{-1}, 24 h)	SI (ΔV = 0–15 m s^{-1}, 24 h)	Unchanged (ΔV = 0 m s^{-1}, 24 h)	SD (ΔV = −15–0 m s^{-1}, 24 h)	RD (ΔV ≤ −15 m s^{-1}, 24 h)
−24 h (t = −24)	3	73	26	16	0
−18 h (t = −18)	4	71	21	23	0
−12 h (t = −12)	4	67	17	31	0
−6 h (t = −6)	5	52	30	33	2
0 h (t = 0)	5	34	21	60	3
6 h (t = +6)	0	13	10	80	15
12 h (t = +12)	0	5	8	73	28
18 h (t = +18)	0	4	2	69	33
24 h (t = +24)	0	4	1	64	29

Note: ΔV is change in the maximum sustained 10-m wind speed.

Table 2. Samples of TCs with intensity change of five categories (RI, SI, unchanged, SD, RD) from 24 h prior to landfall (t = −24), at the time of landfall (t = 0), to 24 h after landfall (t = 24) at every 6 h.

where R_i is each of the individual rain rate estimates and θ_i is the phase angle of the estimate relative to either the storm motion or the VWS vector. The WN-1 asymmetric rainfall component (M_1) can be represented by

$$M_1 = [a_1 \cos(\theta) + b_1 \sin(\theta)] / R. \quad (2)$$

Similarly, the WNs 2–4 asymmetric rainfall components can be calculated respectively.

The shear-relative rainfall asymmetry is defined as the TRMM observed rainfall asymmetry relative to the VWS vector. Similarly, the motion-relative (or coastline-relative) rainfall asymmetries are composite of asymmetric rainfall relative to the motion direction (or coastline orientation) over every TRMM observation. The relative importance of the different factors on the asymmetric rainfall distribution can be characterized.

3. Axisymmetric rainfall

Based on the TRMM 3B42 rainfall data, the amplitude of the WN-0 (axisymmetric component) rainfall is about 0.4–0.6 of the total rainfall in general, though there are some subtle differences in various regions (**Figure 2**). This means that the axisymmetric component could explain about half of the variance of the total rainfall. For a comparison, **Figure 2** also shows the amplitudes of different WN components. The amplitude of the WN-1 component is about 0.2, while the amplitudes of the WN-2, WN-3, and WN-4 components are all less than 0.2 and show smaller and smaller values with the higher WNs. This suggests that on average the axisymmetric component is dominant of the total rainfall and the WN-1 component dominates in the asymmetric rainfall in LTCs. Another feature in **Figure 2** is that there is a large axisymmetric rainfall decrease for the LTCs after landfall in most studied regions over China.

3.1. Axisymmetric rainfall evolution during landfall

The axisymmetric rainfall distributions in LTCs before landfall and after landfall in the five regions of China shown in **Figure 1** are further compared in **Figure 3**. The peak rainfall is about 5.5 mm h^{-1} and located near 60 km from the TC center within 24 h before landfall (**Figure 3a**). After landfall, the peak rain rate is reduced to about 3 mm h^{-1} and located much closer to the TC center (**Figure 3b**). Therefore, the maximum mean axisymmetric rain rate is largely reduced and shifts inward during landfall likely because of the increased surface frictional convergence and the associated eyewall contraction or collapse [66]. Meanwhile, the azimuthally averaged radial profiles of rain rate vary with landfalling regions, which are similar to the radial profiles in the five different ocean basins of the globe as illustrated in [23].

3.2. TC intensity and averaged axisymmetric rainfall

The axisymmetric rainfall is tied to TC intensity on average. **Figure 4** shows the evolutions of TC intensity and the axisymmetric rainfall rate in LTCs in the five regions of China. Generally during

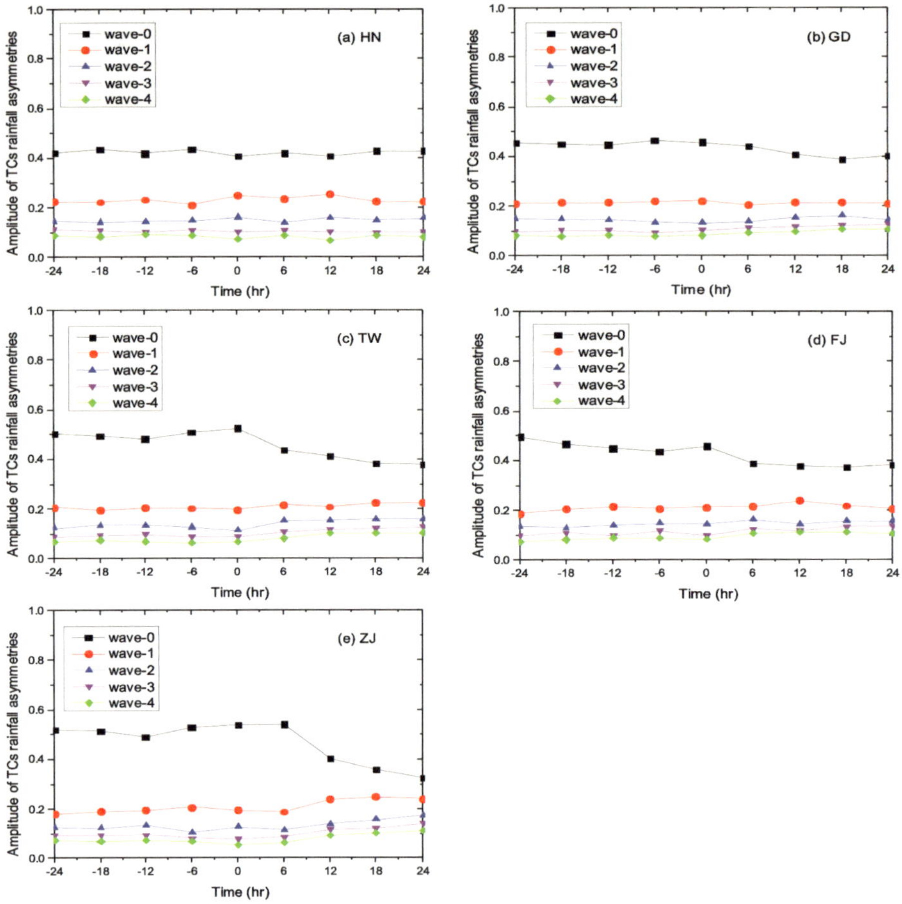

Figure 2. The 6-h amplitudes (×100%) of the WN-0 (axisymmetric) and WN-1, WN-2, WN-3, and WN-4 (asymmetric) components of TC rainfall relative to the total rainfall in regions of (a) HN, (b) GD, (c) TW, (d) FJ, and (e) ZJ from 24 h prior to landfall to 24 h after landfall.

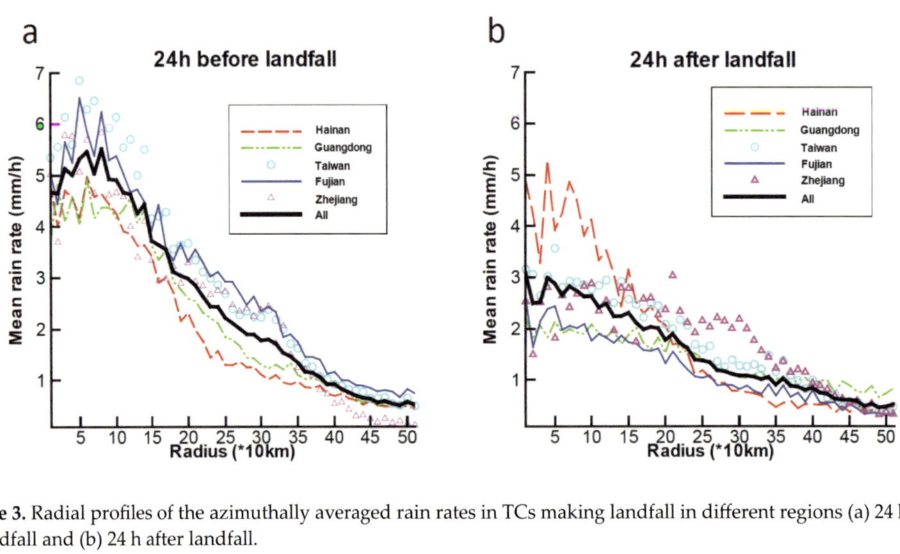

Figure 3. Radial profiles of the azimuthally averaged rain rates in TCs making landfall in different regions (a) 24 h prior to landfall and (b) 24 h after landfall.

24–12 h (t = −24 to −12) prior to landfall, most TCs is still intensifying, but starts to weaken at the time of landfall (t = 0) (**Figure 4a**). During the landfall process (from t = −24 to t = +24), the mean TC maximum wind speed is decreased from 34 to 20 m s^{-1}, namely about 41% decrease in 48 h, while the mean TC minimum sea-level pressure increases from 970 to 992 hPa (**Figure 4b**). During this period, the averaged axisymmetric rain rate shows a very similar variation with the rain rate decreased by about 38% in 48 h (**Figure 4c**). This suggests that the axisymmetric rain rate is positively correlated with TC intensity. In addition, with the decreased intensity of the TCs during landfall, the mean axisymmetric rain rate decreases within a radius of 150 km from the TC center (**Figure 4d**).

3.3. TC intensity and maximum total rain, rain area, and rain rate

Although the axisymmetric rainfall in LTCs is positively correlated with TC intensity, the extreme rainfall parameters in terms of various metrics are not necessarily related perfectly to TC intensity. Here to demonstrate such features, 3 parameters are defined: maximum rain rate (the largest rain rate over all grid points within 500 km of each TC center),

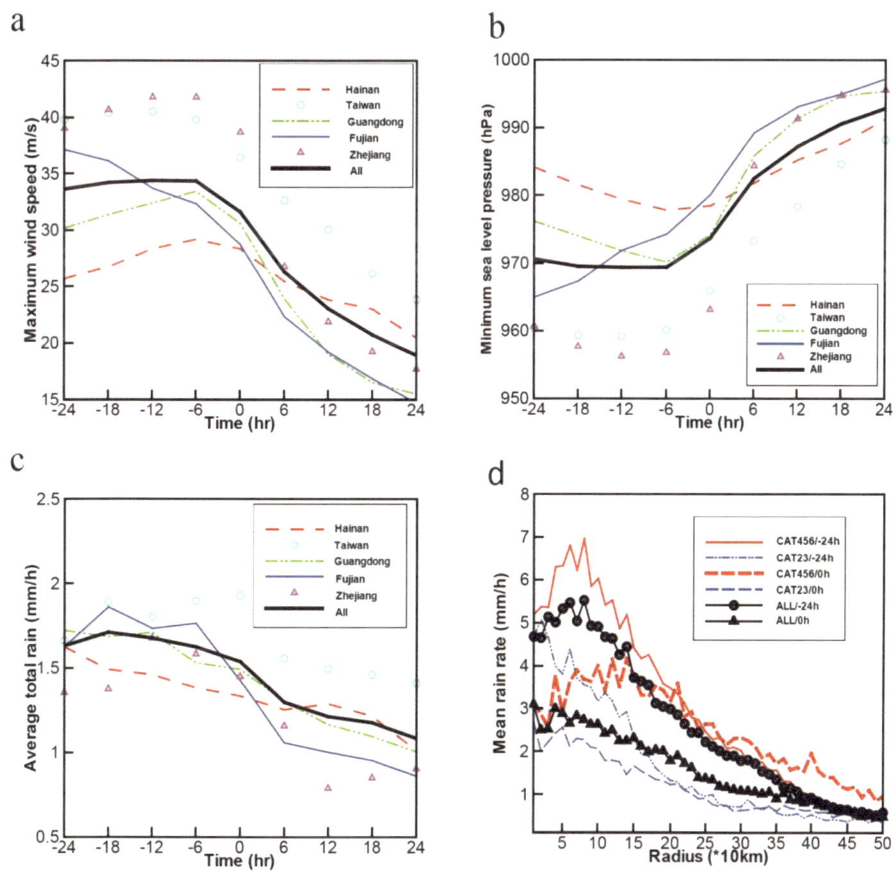

Figure 4. Evolution of rainfall from 24 h prior to landfall to 24 h after landfall at 6-h intervals for (a) the averaged intensity indicated by the maximum sustained 10-m wind speed, (b) the averaged minimum sea-level pressure, and (c) the areal averaged rain rate within 500-km radius of the TC center for all TCs. (d) Radial profiles of the azimuthally averaged rain rates for different TC intensity categories.

maximum total rain (the largest value of accumulated rainfall within 500 km of the TC center among CAT2–6 TC samples), and maximum area of rain (the largest value of fractional rain area relative to the TC area within 500 km radius from the TC center among CAT2–6 TC samples).

As shown in **Figures 5–6**, on average, the CAT6 TCs have the highest rain rate (**Figure 5a**), the highest total rain (**Figure 5b**), and the largest rain area (**Figure 6**), while the CAT2 TCs have the lowest rain rate (**Figure 5a**), the lowest total rain (**Figure 5b**), and the smallest rain area (**Figure 6**).

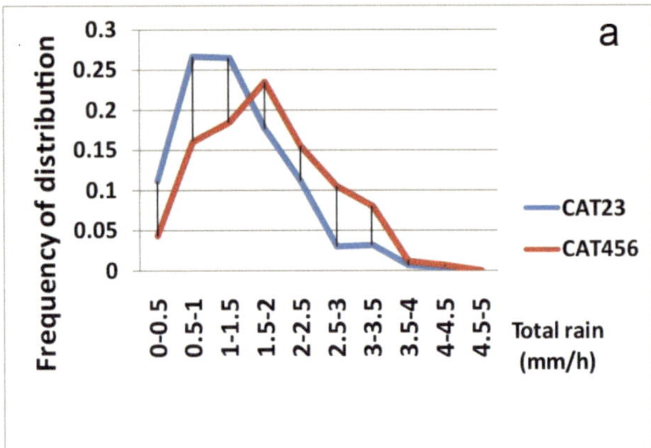

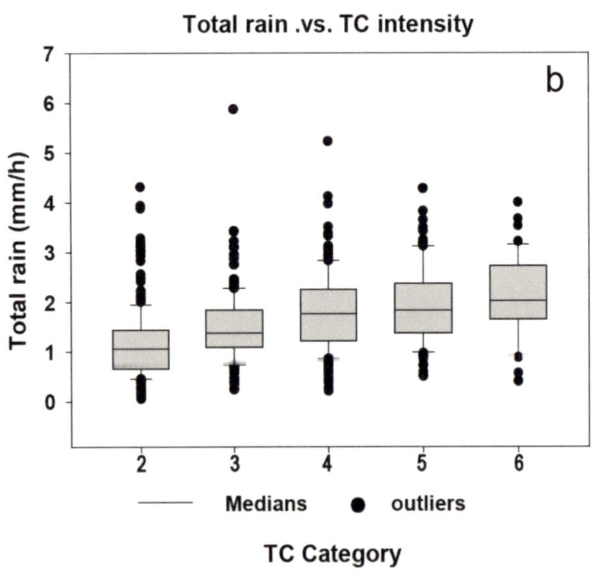

Figure 5. (a) Frequencies of the total rain (the areal averaged rain rate within 500-km of the TC center) for CAT23 and CAT456 TCs. (b) the total rain distributions for TC categories of CAT2–6 from 24 h prior to landfall to 24 h after landfall (medians and outliers in the box plots have the same meaning later).

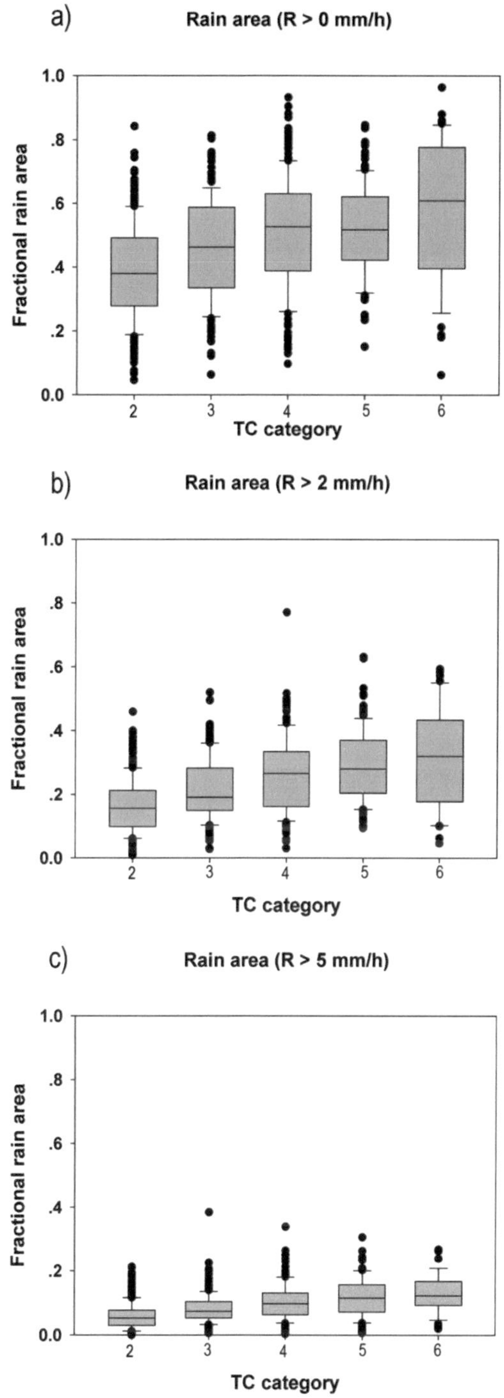

Figure 6. Fractional rain area (×100%) relative to the total area within a 500-km radius from the TC center with different rain rates: (a) R > 0, (b) 2, and (c) 5 mm h^{-1} for TC categories of CAT2–6 from 24 h prior to landfall to 24 h after landfall.

Figure 7a also shows the averaged rain area evolutions of three various rain rates for LTCs of CAT456 and CAT23 during landfall. For both strong and weak TCs, the rain area would shrink to about 10% of the total rain area with the rain rate increasing to being greater than 5 mm h^{-1}. However, the weaker TCs like CAT2 TCs have larger extreme values of the maximum rain

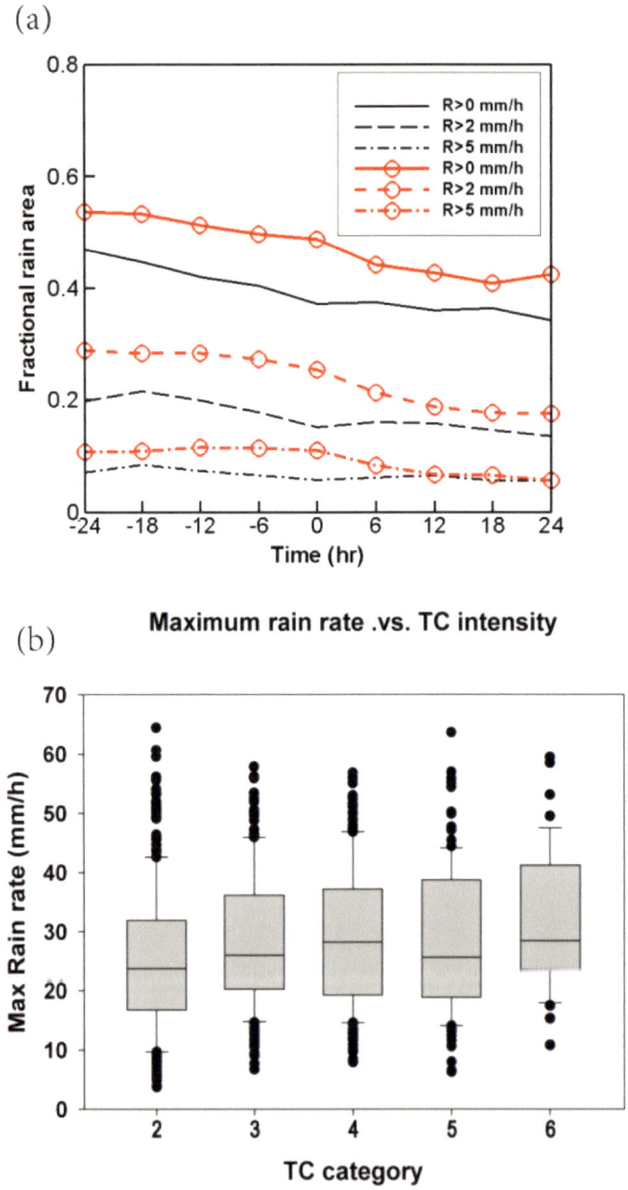

Figure 7. (a) The fractional rain area (×100%) of the total area within a 500-km radius of the TC center with different rain rates >0, 2, 5 mm h^{-1} for CAT456 (with red circles) and CAT23 TCs (without circles) from 24 h prior to landfall to 24 h after landfall. (b) The maximum rain rate (mm h^{-1}) within a 500-km radius of the TC center for TCs of CAT2–6 from 24 h prior to landfall to 24 h after landfall.

rate, total rain, and rain area (**Figures 5b**, **6**, and **7b**). This result thus suggests that the averaged rain features (including rain rate and rain area) are all related to TC intensity, but the extreme rain features (including the maximum rain rate and rain area) are not necessarily so.

3.4. TC intensity change and the axisymmetric rain change

Since the axisymmetric rainfall is related with TC intensity in LTCs, the TC intensity change would be also correlated to the rainfall change. It is shown in **Figure 8a** that on average, the

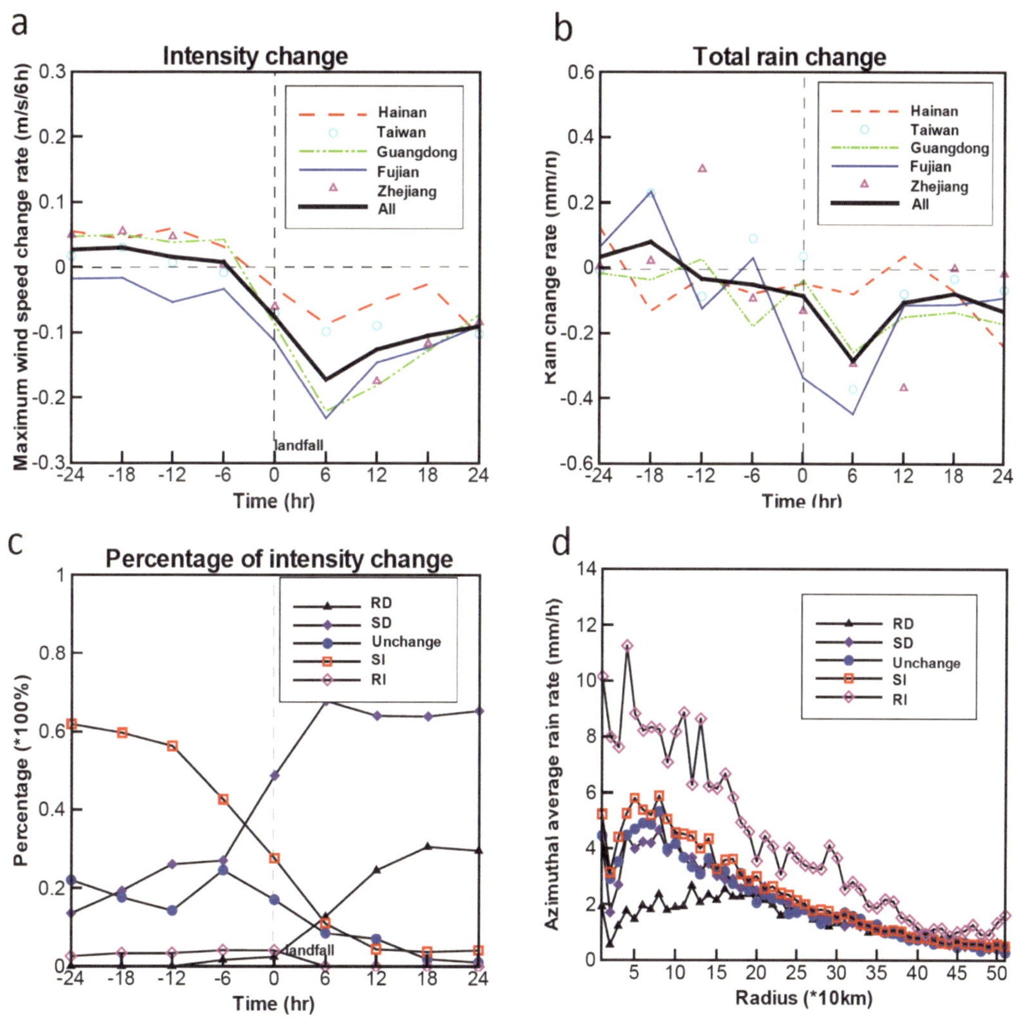

Figure 8. Six-hourly time evolution from 24 h prior to landfall to 24 h after landfall for (a) the rate of averaged TC intensity change (m s^{-1} in 6 h), (b) the rate of change of the areal averaged rain rate (within 500-km radius of the TC center; mm h^{-1}) for all TCs, and (c) the frequency of TCs of rapid decaying (RD), slow decaying (SD), unchanged, slow intensifying (SI), and rapid intensifying (RI). (d) Radial profiles of the azimuthally averaged rain rates for different intensity change groups.

rate of the averaged LTC intensity change is slightly positive or close to zero before landfall, but turns to be negative after landfall. Correspondingly, the rate of the averaged rain rate change presents a similar evolution during landfall (**Figure 8b**). As a result, the correlation coefficient between the LTC intensity change and the axisymmetric rain change reaches 0.77, indicating that they are highly correlated during landfall. **Figure 8c** shows that within 24 h before landfall, 70–90% of LTCs increase or maintain the intensity, while after landfall 20–30% of LTCs have RD process. During the 48-h landfall process, RI rarely occurs in LTCs. **Figure 8d** shows that LTCs of different intensity change categories have different radial distributions in their axisymmetric rain rate. The RI TCs have the highest mean peak axisymmetric rain rate (>13 mm h^{-1}) located at the smallest radius of near 40 km, while the RD TCs have the lowest peak axisymmetric rain rate (about 3 mm h^{-1}) at the radius of about 125 km from TC center. Note also that the total rain rate change from an increase to a decrease for TCs from RI to RD (**Figure 9**).

3.5. TC intensity and the axisymmetric contribution to the total rainfall

Figure 10 compares the axisymmetric contribution to the total rainfall between strong TCs (CAT456) and weak TCs (CAT23). The strong TCs have larger total rainfall in general (**Figure 10a** and **b**), also with a higher contribution from the axisymmetric rainfall component to the total rainfall than the weak TCs (**Figure 10c** and **d**). On average, the weak TCs have lower total rainfall with a larger contribution from the asymmetric components (including WNs 1–4) to the total rainfall.

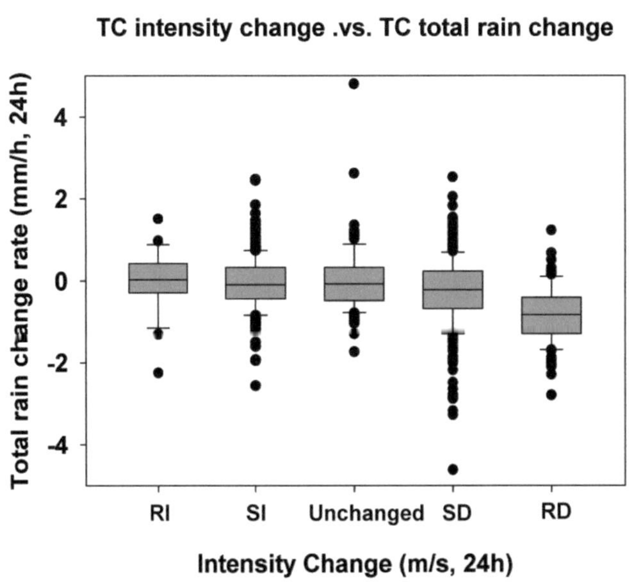

Figure 9. The relationship between different TC intensity change (RI, SI, unchanged, SD, and RD) and total rainfall change from 24 h prior to landfall to 24 h after landfall.

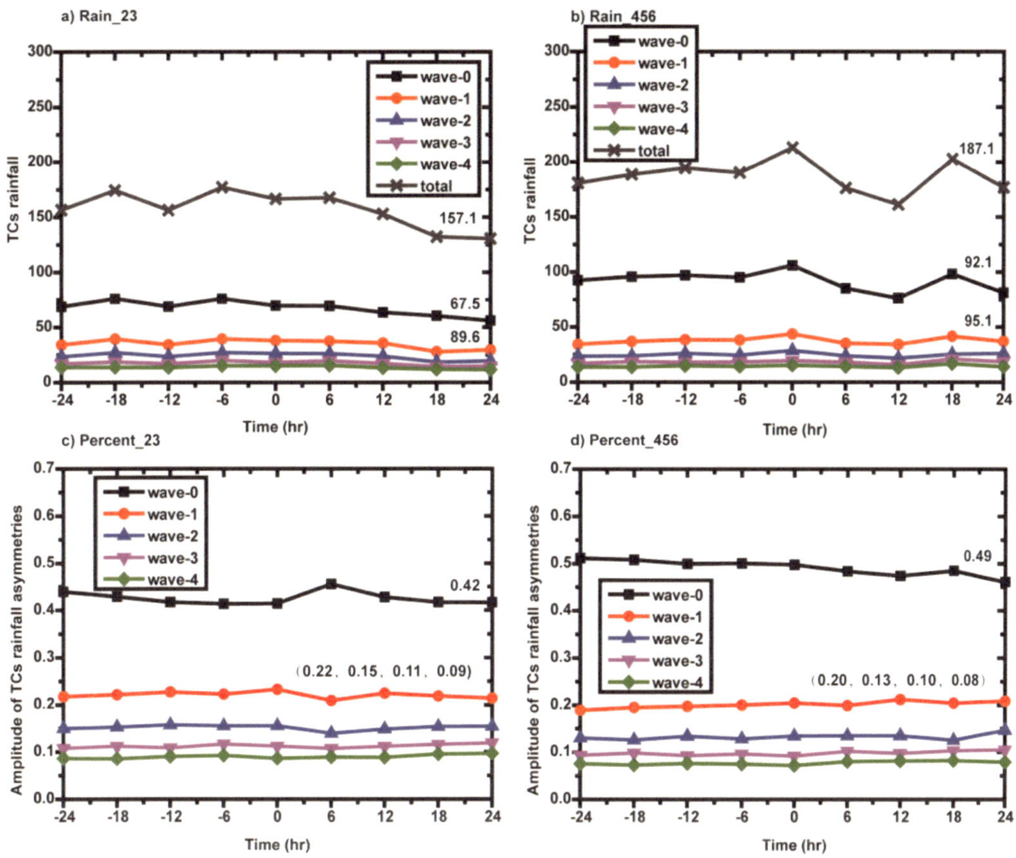

Figure 10. Time evolution of the 6-h WN-0, WN-1, WN-2, WN-3, and WN-4 rainfall components and the total rainfall (mm) from 24 h prior to landfall to 24 h after landfall for (a) CAT23 and (b) CAT456 TCs, and the time integrated mean values of total rainfall, WN-0, and asymmetric rainfall (WNs 1–4) are written above the curves in (a) and (b); (c) and (d) are, respectively, the same as (a) and (b) but for the 6-h amplitudes (×100%) of the WN-0, WN-1, WN-2, WN-3, and WN-4 rainfall components relative to the total rainfall. The time integrated mean values of amplitudes of WN-0, and asymmetric parts (WN-1, WN-2, WN-WN-3, and WN-4, respectively) are written above the curves in (c) and (d).

4. Asymmetric rainfall

Except for the dominant contribution of the axisymmetric component to the TC total rainfall, the asymmetric component, especially that from the WN-1, is also shown to be considerable in previous studies of Yu et al. [25, 26]. The WN-1 rainfall asymmetry in LTCs in the five regions of China (**Figure 11**) shows that the maximum rainfall was generally located in the south of TC center before landfall and slightly rotates cyclonically from HN, GD, TW, FJ to ZJ, namely, from South China to East China. This cyclonical rotation still exists after landfall with the maximum rainfall cyclonically rotated to the northeast in LTCs in ZJ of East China. The effects of some important factors including VWS, TC motion, intensity, and coastline on the LTC asymmetric rainfall distribution are also discussed below.

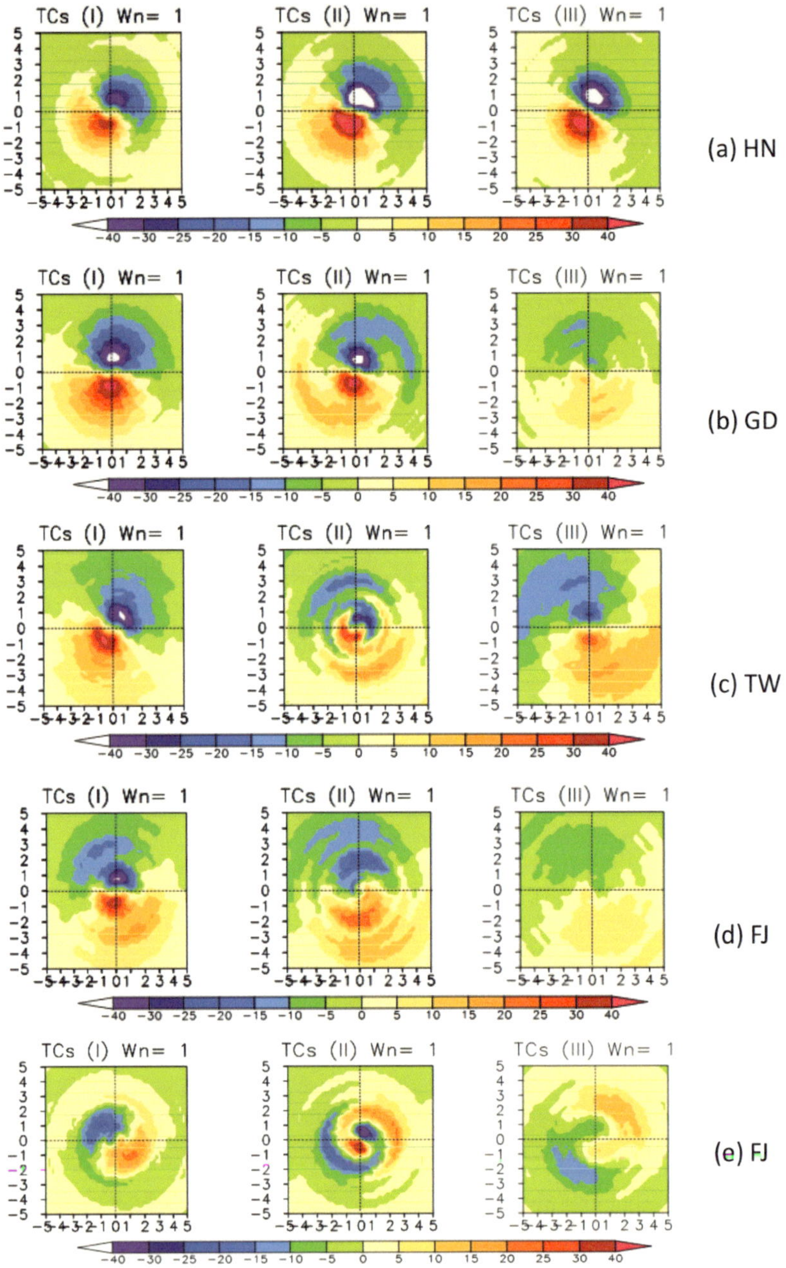

Figure 11. The WN-1 rainfall asymmetry (unit: mm) as a function of the distance from the TC center for TCs making landfall in different regions: (a) Hainan, (b) Guangdong, (c) Taiwan, (d) Fujian, and (e) Zhejiang in three stages. X and Y axes are the distance (degree) from the TC center (origins). Stage (I) is 24 h prior to landfall, stage (II) is at the time of landfall, and stage (III) is 24 h after landfall. The color scale indicates the amplitude of the asymmetry relative to the distance from the TC center.

4.1. Effect of vertical wind shear

Most LTCs in South China are still far away from the midlatitude westerlies, and therefore, in general no direct impacts of westerly troughs on LTCs occur. However, the LTCs in East

China are often close to the westerly systems (troughs and ridges) from the north, and thus they are more frequently and directly affected by the midlatitude westerlies so that the westerly trough would impinge into the LTC circulations. As a result, the high VWS, the water vapor flux channel, and the accompanied baroclinic zone often impose significant impact on rainfall distribution in the LTCs in East China, especially after landfall in Yu et al. [25].

Figure 12 shows the changes of the environmental VWS during 24 h before landfall to 24 h after landfall. It shows that prior to landfall, the LTCs are affected by easterly VWS for regions of HN, GD, TW, and FJ, while westerly VWS for ZJ region (as shown by green curves in **Figure 12**). After landfall, the averaged easterly VWS is generally decreased with time during landfall except for a strengthening of the averaged VWS and a variation from westerly VWS to southwesterly VWS for ZJ region (as shown by red curves in **Figure 12**). Therefore, there exists a cyclonic rotation of VWS for the LTCs from South China to East China both prior to and after landfall.

The shear-relative WN-1 rainfall distribution in TCs making landfall in the five regions from South China to East China confirms that the asymmetric rainfall maximum in LTCs in all five regions is mainly located downshear to downshear-left whenever 24 h prior to landfall, or at the time of landfall, or 24 h after landfall (facing down the VWS vector aligned with the positive Y-axis in **Figure 13**). The cyclonic rotations of the rainfall maximum in the LTCs in the five regions are thus closely related to the cyclonic rotation of VWS. This indicates that VWS is a dominant factor affecting the asymmetric rainfall distribution in TCs; regardless the TCs are over open oceans or making landfall.

4.2. Effects of TC intensity

When the TCs are separated into two groups of CAT23 (weak TCs) and CAT456 (strong TCs) based on their intensity, we can find that in both CAT456 and CAT23 TCs, the maximum rainfall is always located downshear to downshear-left during landfall from t = −24 to t = 24 (**Figure 14**). However, the mean VWS magnitude for CAT23 TCs is 8.6 m s^{-1}, which is larger than 6.0 m s^{-1} for CAT456 TCs. The rainfall maximum in CAT456 TCs is a little bit cyclonically rotated from the VWS vector relative to that in CAT23 TCs. This might be related to the differences in both the VWS magnitude and the TC intensity. First, for strong VWS and weak TCs, the location of rainfall maximum is often located in downshear to downshear-left quadrant when facing downshear direction. But for weak VWS and strong TCs, the rainfall maximum could rotate further cyclonically to the downshear-left or even to the left of the shear vector. This indicates the relationship between VWS and rainfall asymmetry is determined not only the magnitude of the VWS but also the TC intensity.

4.3. Effect of TC motion

During the landfall period from 24 h prior, to landfall, and to 24 h after landfall, the motion-relative WN-1 rainfall maximum in LTCs in the five regions of China shows large variability (facing down the TC motion vector aligned with the positive Y-axis in **Figure 15**). The rainfall maximum for LTCs in HN is often to the left of the motion vector, while for LTCs in FJ the rainfall maximum occurs in the rear side of the motion vector. This indicates that the rainfall asymmetry may not be predominantly controlled by the TC motion vector for LTCs over China.

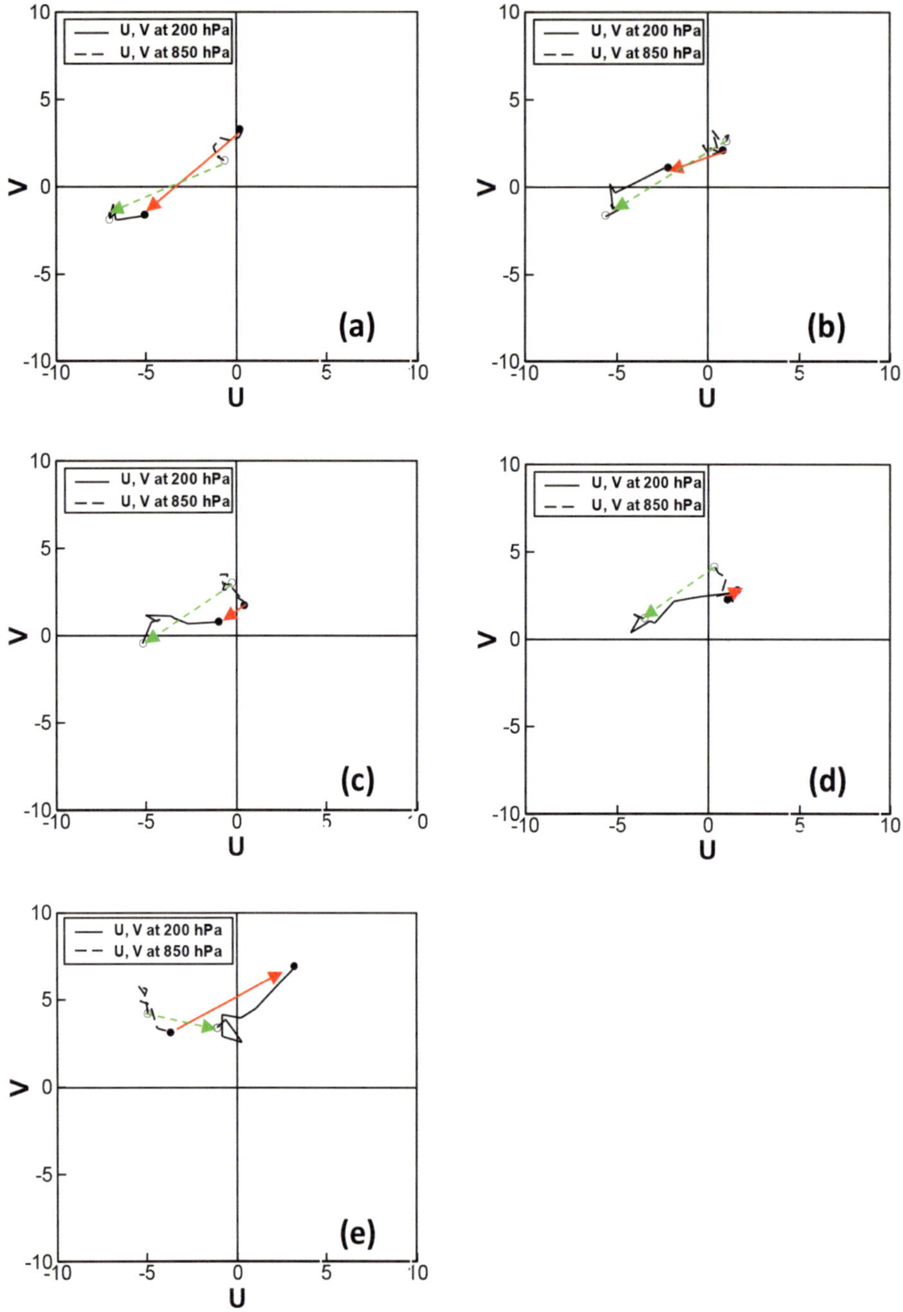

Figure 12. Changes in the averaged zonal wind (U, unit: m s^{-1}) and meridional wind (V, unit: m s^{-1}) within 500 km of the TC center for TCs making landfall in (a) Hainan, (b) Guangdong, (c) Taiwan, (d) Fujian, and (e) Zhejiang. The open and solid circles represent the 9 points (at every 6 h) starting at 24 h prior to landfall and ending at 24 h after landfall, respectively. The black solid lines are wind changes at 200 hPa, and the black long dashed lines are wind changes at 850 hPa. The green (red) dash lines represent the vertical wind shear at 24 h prior to (24 h after) landfall.

http://dx.doi.org/10.5772/intechopen.75910

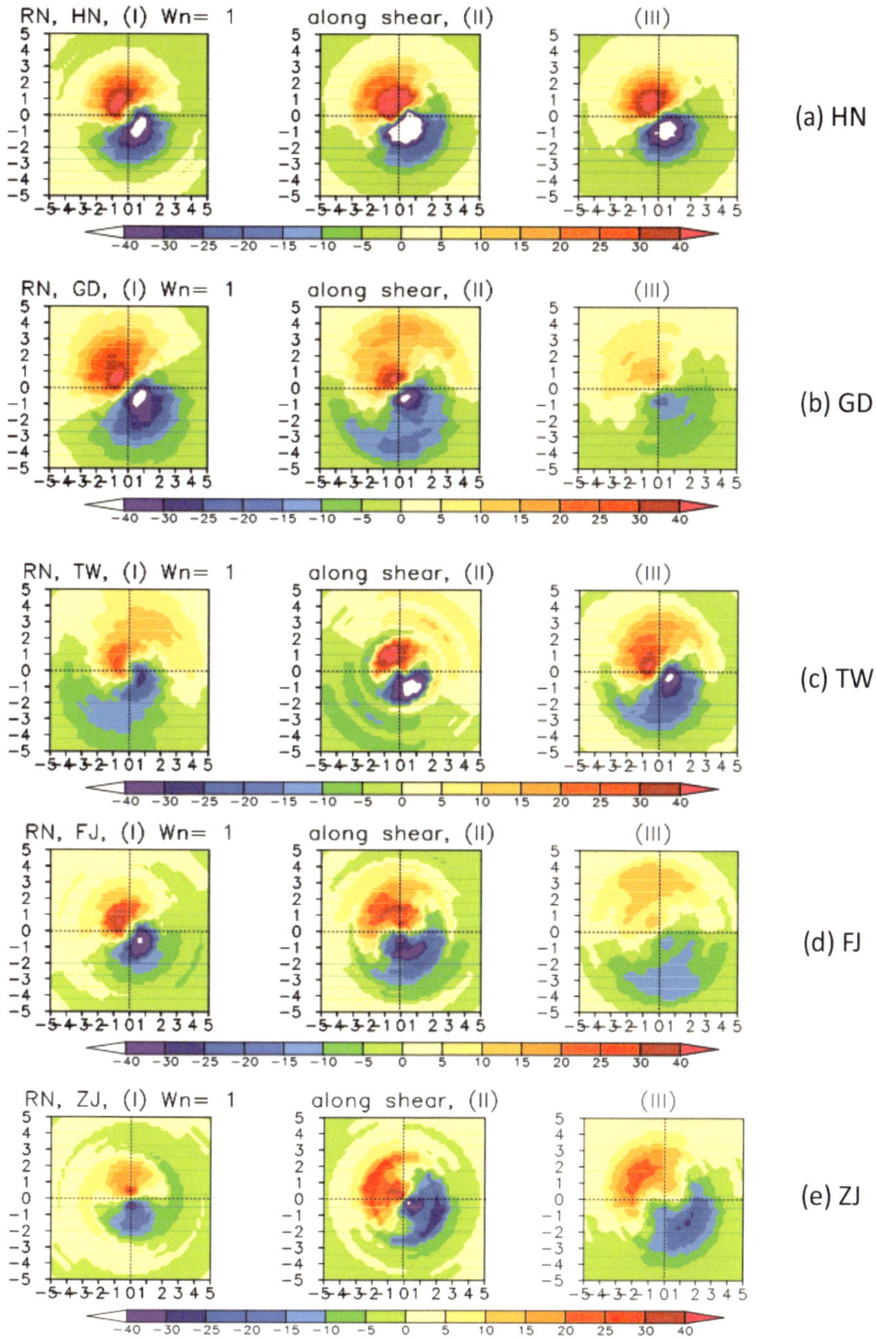

Figure 13. Same as in **Figure 11** but for the WN-1 rainfall asymmetry (unit: mm) relative to the VWS. The shear vector is aligned with the positive Y-axis (upward). X and Y axes are distance (in degree) from the TC center (origins). (I), (II), and (III) are 24 h prior to landfall, at the time of landfall, and 24 h after landfall, respectively. The color scale indicates the amplitude of the asymmetry relative to the VWS.

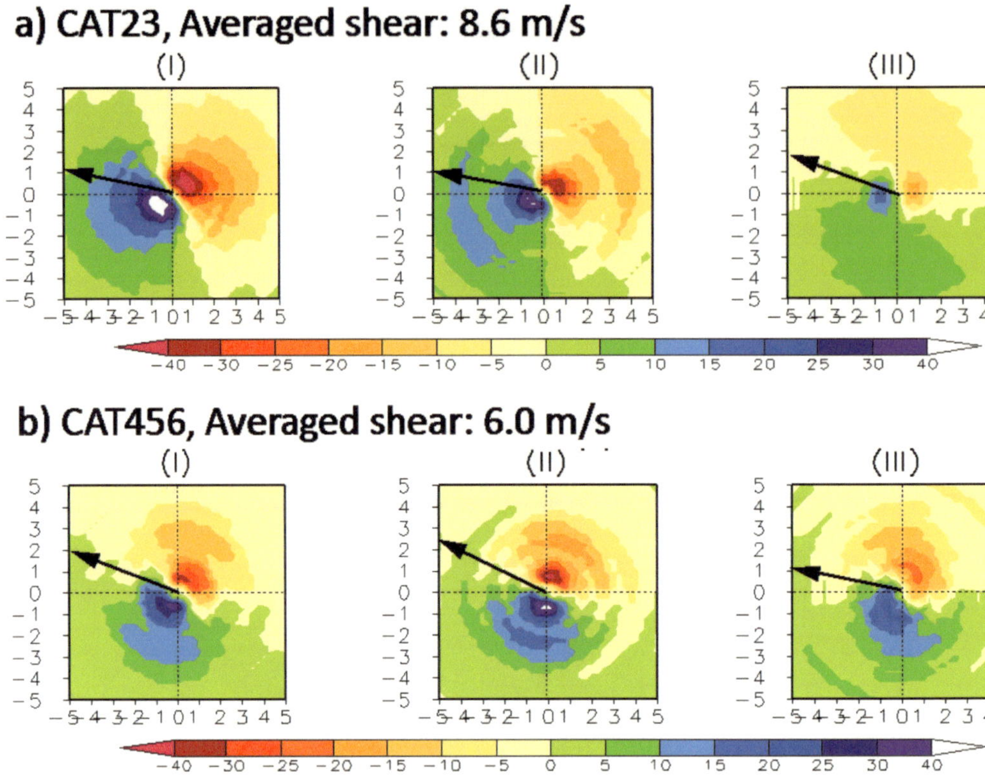

Figure 14. The WN-1 rainfall asymmetry (shaded, mm) for (a) CAT23 and (b) CAT456 TCs. The solid arrows denote the averaged VWS vector. X and Y axes are distances in longitude and latitude (in degree) from the TC center (origin). (I), (II), and (III) are 24 h prior to landfall, at the time of landfall, and 24 h after landfall.

In addition to the motion direction, the motion speed shows inconsistent relationship with rainfall asymmetry as well. **Figure 16** shows rainfall asymmetries in TCs of three different translational speeds: less than 12, 12–20 km h^{-1}, and larger than 20 km h^{-1}. We can see that regardless of different motion speeds, the maximum rainfall is predominantly located in the left and even rear-right quadrants when facing down the TC motion vector. This means that the TC motion speed has no obvious impacts on the rainfall asymmetric distribution in LTCs over China. This may partly suggest that the effects of storm motion on the asymmetric rainfall distribution in the studied regions are dominated by the effects of environmental VWS.

4.4. Effects of coastlines

In contrast to the negligible effects by LTC motion, the WN-1 rainfall maximum is strongly affected by the coastlines, especially in the LTC outer core region and at the time of landfall. **Figure 17** shows the WN-1 rainfall asymmetries for LTCs with different magnitudes of VWS (less than 5, 5–7.5 m s^{-1}, and larger than 7.5 m s^{-1}, respectively). In the inner core region (within a radius of 2° of latitude), the maximum rainfall in LTCs is generally located offshore and downshear-left under all VWS conditions during landfall. However, in the outer core

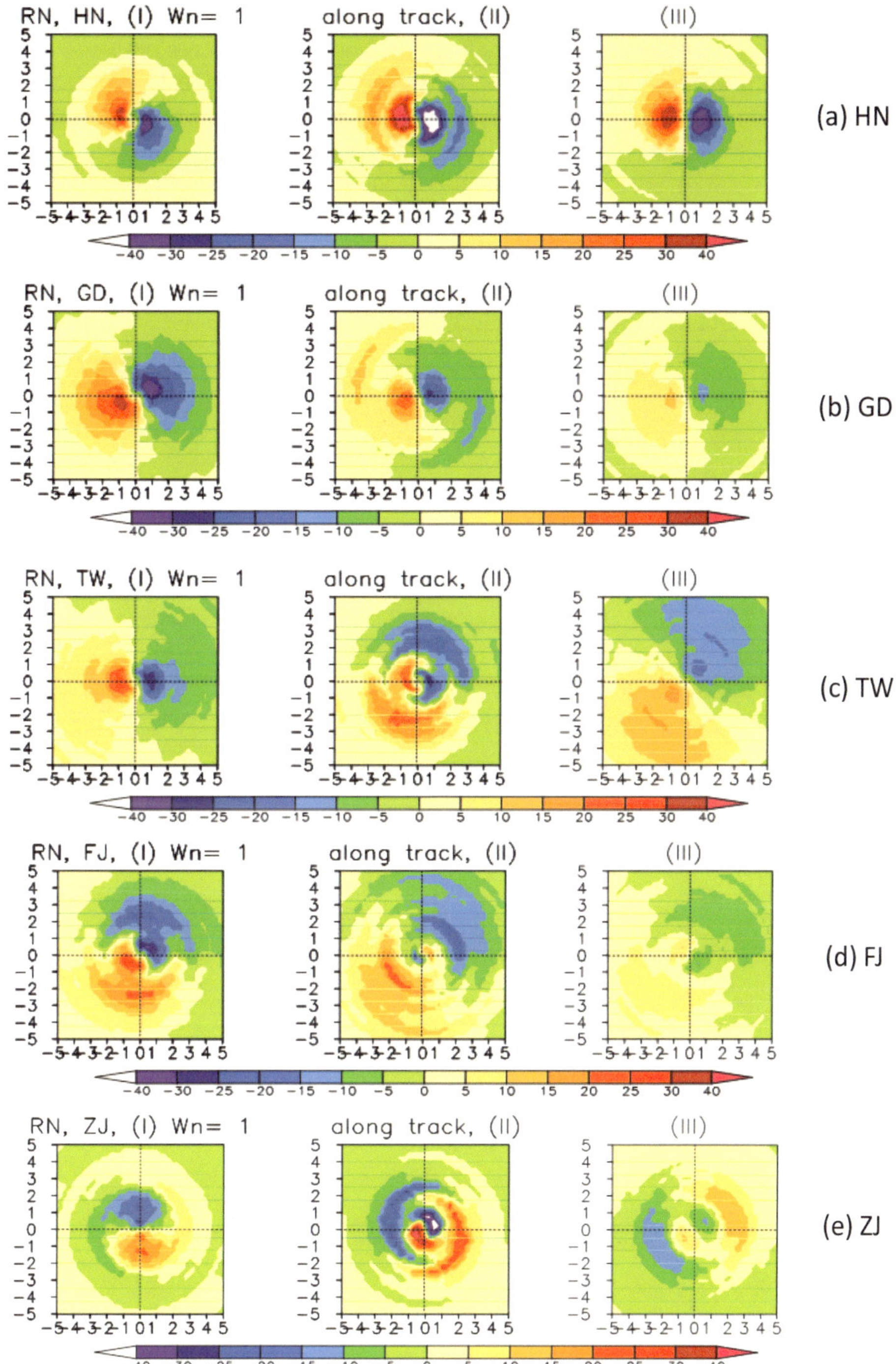

Figure 15. Same as in **Figure 13** but for the WN-1 rainfall asymmetry (unit: mm) relative to the TC motion. The motion vector is aligned with the positive Y-axis (upward). X and Y axes are distance (degree) from the TC center (origins). (I), (II), and (III) are 24 h prior to landfall, at the time of landfall, and 24 h after landfall. The color scale indicates the amplitude of the asymmetry relative to the TC motion.

region (between 2 and 5° of latitude from the TC center), the asymmetric rainfall maximum is rotated cyclonically from offshore gradually to the onshore in TCs under decreasing environmental effects from strong VWS to weak VWS (namely, from larger than 7.5 m s^{-1} to less than 5 m s^{-1}). This result is consistent with that previously reported in a numerical study of Xu et al. [45], which showed that the land-sea surface contrast could produce an increase in the rainfall percentage onshore in a LTC. Results in Yu et al. [26] showed that when the VWS is very weak, the land-sea contrast could predominantly contribute to the rainfall asymmetry. Nevertheless, the situation could become more complicated under weak VWS, because other factors including the coastline and storm motion might play their roles together to determine the asymmetric rainfall distribution in LTCs, particularly in the TC outer core and at time of landfall.

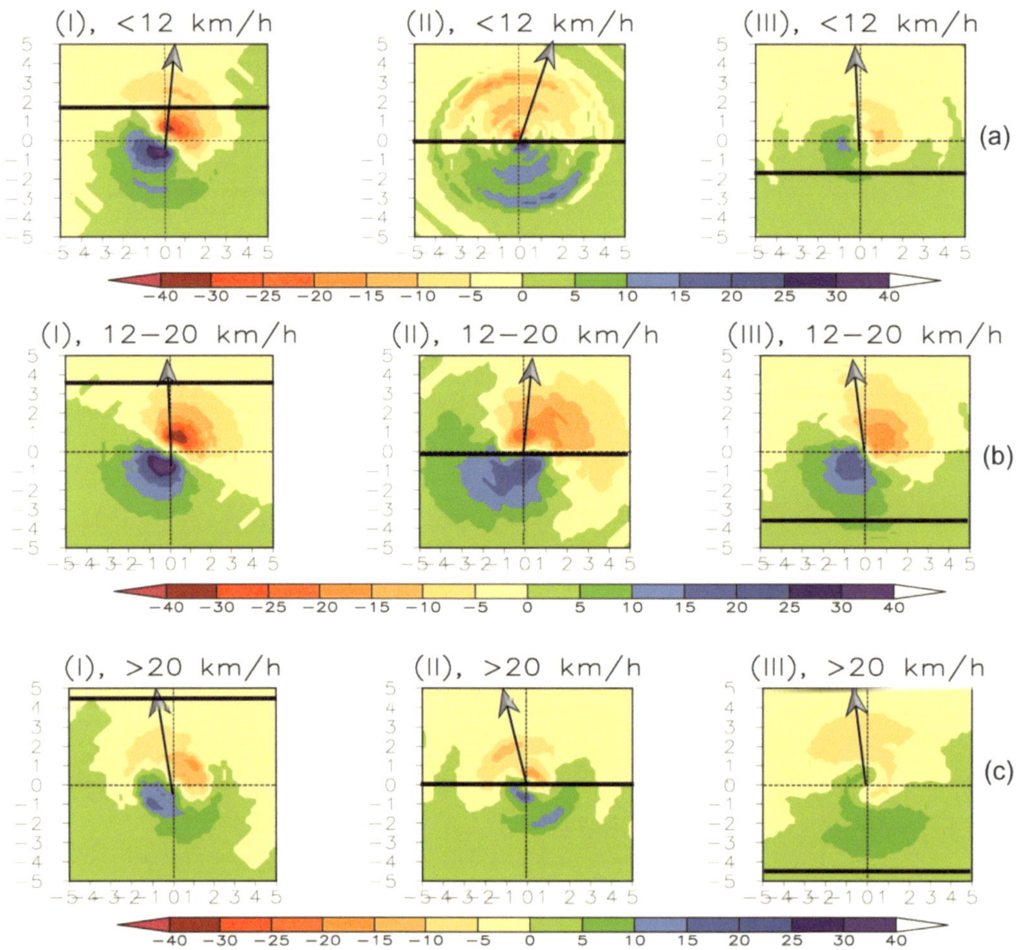

Figure 16. The WN-1 rainfall asymmetry (shaded, mm) relative to the coastlines for TCs with different translational speeds: (a) <12 km h^{-1}, (b) 12–20 km h^{-1}, and (c) > 20 km h^{-1}. The coastline is aligned with the positive X-axis to the right (shown by the black solid lines) and the arrows denote the mean motion direction.

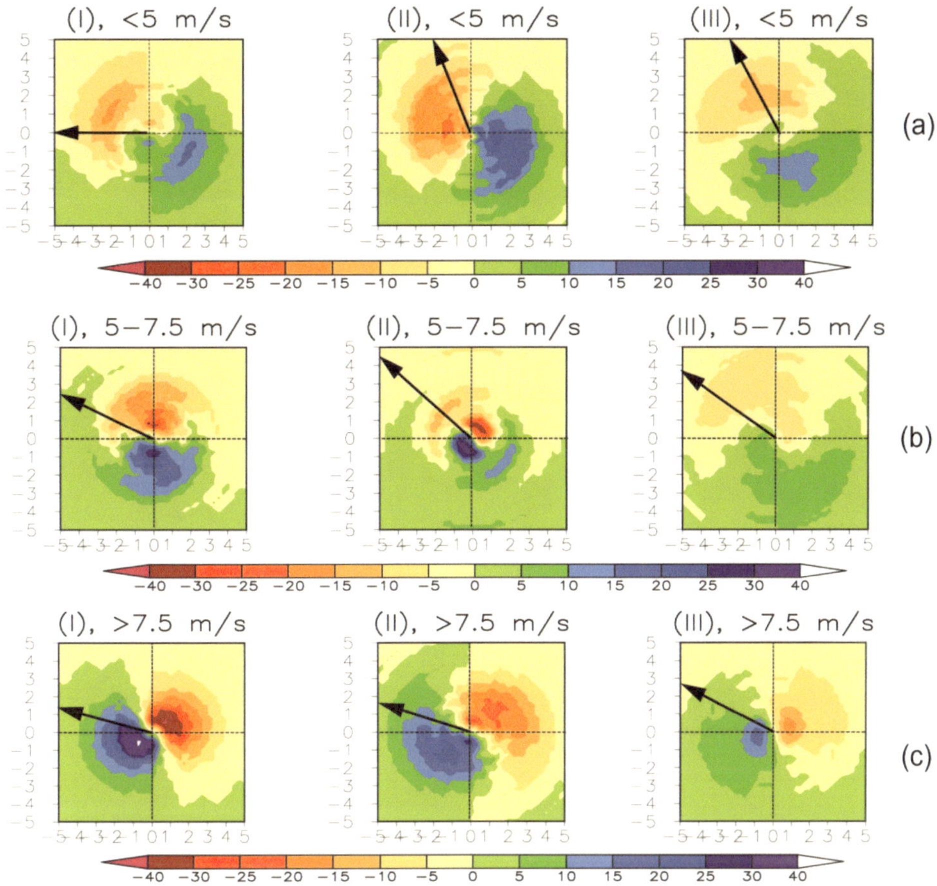

Figure 17. Same as in **Figure 16** but for TCs with different VWS magnitudes: (a) <5 m s^{-1}, (b) 5–7.5 m s^{-1}, and (c) >7.5 m s^{-1}. The solid arrows denote the averaged VWS vectors.

5. Case studies

Since LTC rainfall involves complicated process, such as multi-scale interactions with various scale motions in the environment, coastline, orography, as well as the structure and intensity of the storm itself, detailed case studies can help to deepen our understanding of the various processes that are essential to the rainfall distribution in LTCs. Here two recent case studies are briefly discussed as an illustration how important the cold pool dynamics and the baroclinic interactions are to extreme rainfall associated with LTCs in China.

5.1. Typhoon Utor (2013)

Typhoon Utor (2013) is one of the strongest TCs over the western North Pacific in 2013. It made landfall in Yangjiang, Guangdong Province, China, on 14 August and persisted for about 4 days inland over South China and brought extensive precipitation to most regions of South China.

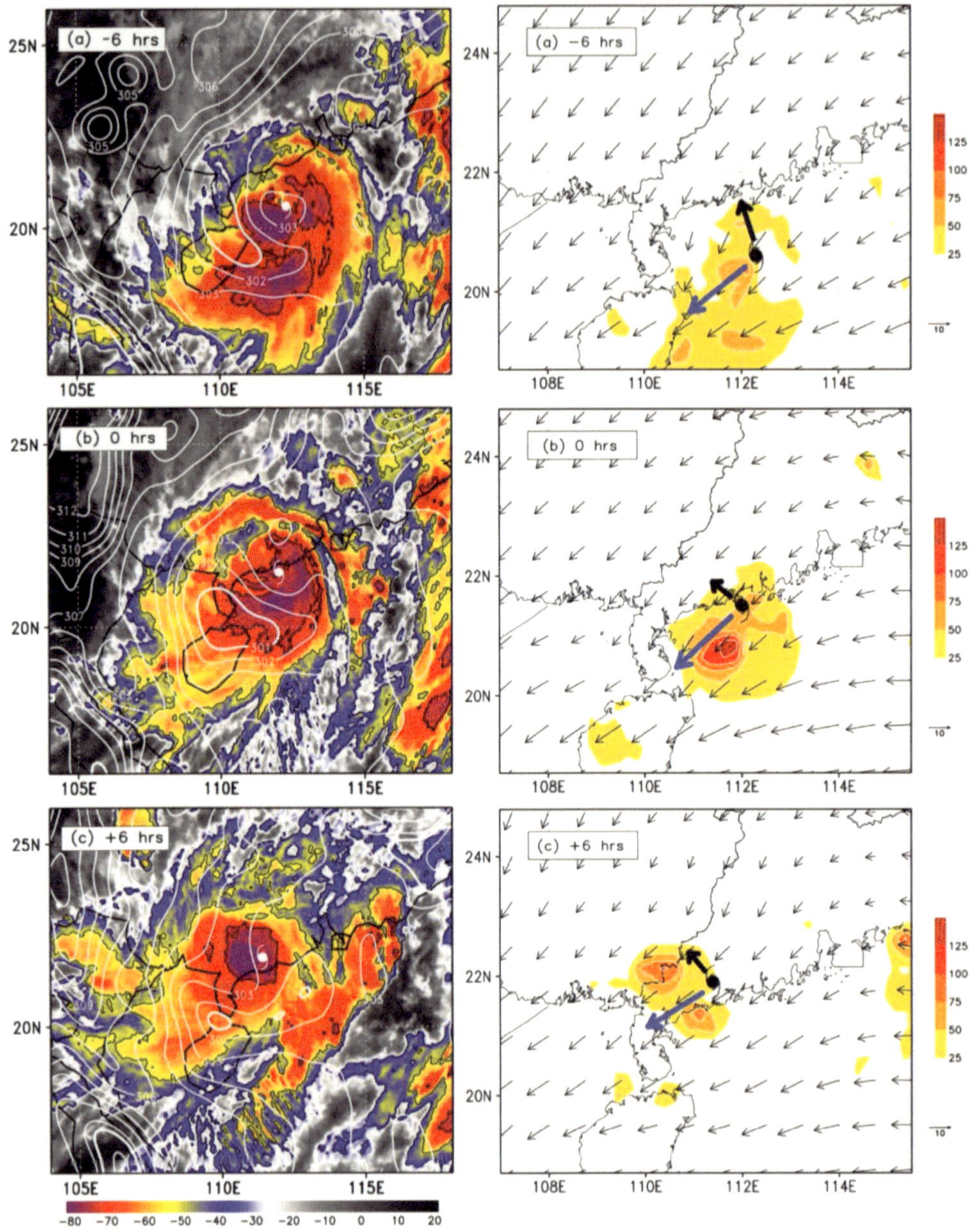

Figure 18. (Left) Potential temperature at 925 hPa (white contours with 1 K interval) and MTSAT IR cloud images (shading, K) at (a) 0000 UTC, (b) 0600 UTC, and (c) 1200 UTC 14 August 2013. The solid thick white contour of 301 K is used to denote the cold pool mentioned in the text. The TC center at the given time is marked by the typhoon symbol in each panel. (Right) Distributions of the environmental VWS vectors (m s^{-1}) and TRMM-3B42 6-h rainfall (mm) during the landfall of typhoon Utor in the first rainfall episode at (a) 6-h prior to landfall at 0000 UTC 14, (b) at the time of landfall at 0600 UTC 14, and (c) 6-h after landfall at 1200 UTC 14 August. Thick black vector denotes the moving direction of Typhoon Utor in the following 6 h in each panel. Blue arrows denote the averaged environmental VWS vector.

The rainfall distribution associated with Utor during landfall presented unique features [46, 47]. Meng and Wang [46, 47] used the ERA-Interim (European Centre for Medium-Range Weather Forecasting Interim reanalysis) data and the hourly surface rainfall observations to investigate the physical processes leading to large rainfall asymmetric distribution of Utor during its landfall. They found that the rainfall asymmetry was initially triggered by the effect of the environmental VWS, but enhanced by the cold pool dynamics which was triggered by the inland northwest dry air intrusion into Utor (**Figure 18a**). This cold pool enhancement process was understood based on the frontogenesis and quasi-balanced and ageostrophic Q-vector diagnosis (**Figure 18b**).

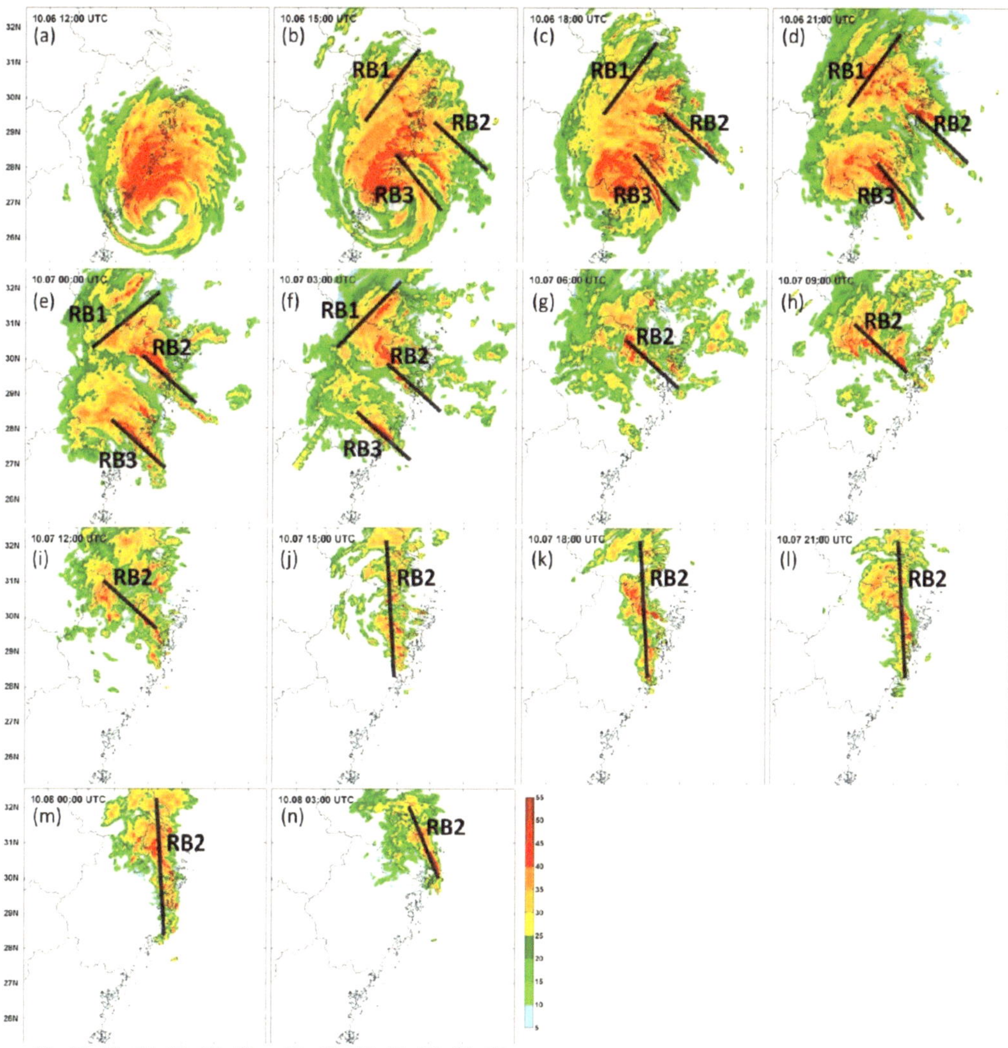

Figure 19. Observed composite radar reflectivity every 3 h from (a) 1200 UTC 6 Oct to (n) 0300 UTC 8 Oct 2013. RBs 1, 2, and 3 indicate rainbands associated with heavy rain. The locations of the four radar stations (Wenzhou, WZ; Ningbo, NB; Hangzhou, HZ; and Shanghai, SH) are shown in (a).

5.2. Typhoon Fitow (2013)

Typhoon Fitow (2013) is the strongest typhoon that made landfall over China in October since 1949. It caused severe damages in Zhejiang province due to its strong winds and heavy rainfall, with the minimum central sea-level pressure of 955 hPa and the maximum sustained 10-m wind speed of 42 m s^{-1} at landfall. The maximum gust wind speed of 76.1 m s^{-1} was observed at Shipingshan station (in the southeast coast of Zhejiang Province), which set a new record for LTCs over China. Fitow brought the strongest large-area 24-h precipitation in Zhejiang Province (100,000 km^2) as large as 137.5 mm on 07 October 2013. Consequently, Fitow resulted in severe urban water logging, flooding, debris flow disasters, and thus a direct economic loss of about 4.5 billion US dollars in Zhejiang Province. It also brought more than 1000 mm rainfall near Shanghai area in China, which was found to be related the cold air intrusion from northwest inland as shown in Bao et al. [67]. As shown in **Figure 19**, the rainfall systems included two parts: one was an outward-spiraling rainband, and the other was a frontal cloud system caused by coastal frontogenesis. Fitow was affected by a cold dry air intrusion from the north after landfall, which limited the rainfall in its inner core. However, an extended eastern warm moist air met the cold air from the north, leading to frontogenesis and heavy rainfall near Shanghai.

6. Discussions and remaining issues

Yu et al. [68] have presented verification of objective 0–6, 0–24, 24–48, 48–72, and 0–72 h rainfall forecasts for 25 LTCs and 133 operational numerical predictions from the tropical cyclone version of the Australian Community Climate and Earth System Simulator (ACCESS-TC). The contiguous rain area (CRA) method was used to diagnose the origin of systematic errors via adjustment of the forecast rain field by displacement, rotation, volume, and pattern. Mean track and intensity errors at 48 h for the sample are, respectively, 180 km and 10 knots. These represent a quite skillful level of performance of the ACCESS-TC. The mean values of equitable threat score, probability of detection, and false alarm ratio for the 30-mm isohyet for the unadjusted forecasts at 0–6 h are (0.22, 0.56, 0.66) (**Figure 20**). Unsurprisingly skill of the 24-h accumulated rainfall forecasts is the highest (0.36, 0.63, 0.40) for the 0–24-h forecast and then declines to (0.20, 0.42, 0.60) for the 72-h forecasts. Forecast skill also declines with rainfall amount.

With the use of the CRA method, skills show an improvement by about 15%, which are mostly associated with rainfall patterns, followed by displacement errors, particularly for very heavy rainfall. This suggests that rainfall prediction will continue to be improved with improving track prediction. However, further analysis showed that though the TC track and intensity is well forecasted, the rain distribution and structure is still a big challenge to the current numerical models, especially for the extremely heavy rainfall (>250 mm day^{-1}). It indicates that more work is needed to improve the initialization and prediction of TC structure. In particular, the TC size is a very important factor that could considerably affect the rainfall distribution in LTCs. However, this has not been studied so far mainly because the TC size information in the current TC best-track data is very limited for LTCs. In future studies, the

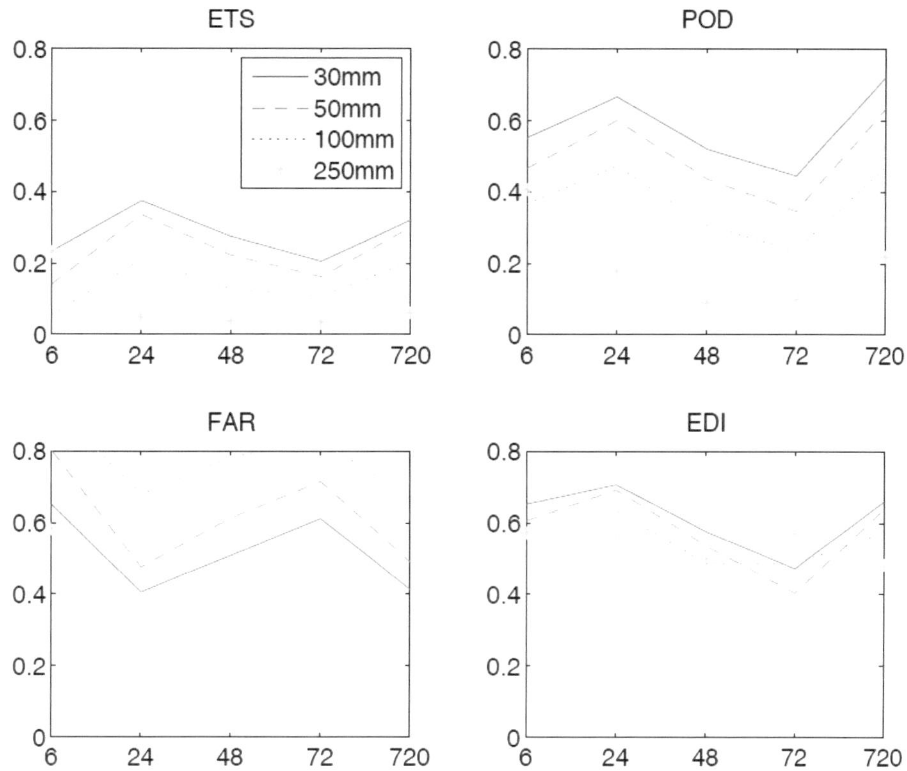

Figure 20. Conventional verification results of ETS, POD, FAR, and EDI for 30, 50, 100, and 250 mm rainfall thresholds. X-axis shows the accumulated rain forecast lead time of 0–6, 0–24, 24–48, 48–72, and 0–72 h, respectively. Y-axis shows the skill scores of the forecasts.

geostationary imagery or other data could be used to estimate the TC size parameter so that the effect of TC size on rainfall distribution of LTCs could be examined.

In fact, except for the limited ability of numerical forecast models, the observational capability of rainfall in LTCs is not good enough. As introduced in Section 2, compared with the surface radar and gauge data, the satellite rainfall estimates have relatively better spatial and temporal coverages, and therefore, the satellite-retrieved rainfall datasets (such as TRMM 3B42) have been applied extensively in the studies of rainfall in LTCs. However, the satellite quantitative precipitation estimates for LTCs still need large improvements because of their existing limitations, e.g., they would underestimate the heavy rain rates and the maximum rain rates. Therefore, the composites of rainfall characteristics were often analyzed and the uncertainties were usually represented using the box-plot analysis method in previous studies in Yu et al. [25, 26]. Though this would not significantly affect the major conclusions drawn in those previous studies, we still need to be aware of the limitations of data used. Therefore, future studies are needed to verify previous findings when better rainfall datasets become available.

Most of previous studies have focused mainly on the overall effects of environmental VWS, TC intensity and motion, and coastlines on the LTCs rainfall distribution and provide some basic background and preliminary understanding to the LTC rainfall. However, it has shown that TC intensity does not have a robust relationship with the extreme rain metrics (including the maximum values of rain rate, rain area and total rain, and even the maximum rainfall location). Weak TCs may even have the maximum rain rates larger than strong TCs after landfall. Therefore, many unresolved scientific questions still remain especially for the extreme rainfall process in LTCs. Further work needs to pay more attention on the extreme rainfall mechanisms in LTCs.

Furthermore, the rainfall processes in LTCs are often very complicated and thus involve the effects of many factors and multi-scale interactions. Most of our current results and understanding are based on dynamical mechanisms, such as the VWS, TC translation, coastline, etc. So far few studies have paid attention to other possible effects, such as the thermodynamic effects (e.g., land-sea surface temperature contrast, differential radiation effects over land and ocean, effect of urbanization, etc.), even the cloud microphysical processes, and the possible effects of aerosols. Therefore, most of our current understanding and explanations are incomplete. In this regard, studies on mechanisms responsible for the rainfall distribution and its changes need to be further enhanced in the future. Particularly, with the development of both new observing systems and advanced numerical weather prediction models, some important physical processes, such as the boundary layer process and cloud microphysics, could be investigated and understood more comprehensively in the near future.

7. Concluding remarks

In this chapter, presented is the rainfall distribution in LTCs over China, including the axisymmetric and asymmetric distributions and the main control parameters, such as TC intensity, environmental VWS, TC motion, and coastline. The amplitudes of the WN 0–4 rainfall components have different evolutions during landfall among different regions. The WN-0, namely the axisymmetric rainfall component, accounts for about 50% of the total rainfall in LTCs in most studied areas and often decreases significantly after landfall.

On average, the axisymmetric rainfall is closely related to the TC intensity. The stronger TCs have larger averaged rainfall rate, larger mean peak axisymmetric rainfall rate, larger averaged total rainfall, larger averaged rainfall area, larger ratio of the axisymmetric component relative to total rainfall, but lower amplitudes of the WN-1, WN-2, WN-3, and WN-4 rainfall components relative to the total rainfall. The evolution of rainfall distribution is closely related to the changes of TC intensity from 24 h prior to landfall to 24 h after landfall. In the different intensity change categories, the RD LTCs show the most rapid decrease in both the total rainfall and the axisymmetric rainfall component.

In contrast, the extreme values of LTC rainfall (including the maximum rain rate, the maximum rain area, and the maximum total rainfall) are not highly related to TC intensity. In fact, weak storms may produce a greater maximum rain rate rather than strong TCs. This may be due to the effects of many other factors, such as environmental VWS and strong interactions with other synoptic weather systems and/or topography. The maximum asymmetric rainfall in LTCs is usually located downshear and to its left, but more cyclonically downwind in strong LTCs such as CAT456, likely due to the relatively weaker VWS and the larger azimuthal advection by the stronger cyclonic circulation. The cyclonic rotation of the location of the LTC rainfall maxima also exists from South China to East China (HN, GD, Taiwan, Fujian, and Zhejiang), namely from the largest rainfall in the southwest and south quadrants in HN, GD, TW, and FJ, to the southeast quadrant in ZJ prior to landfall, and from the southwest to the northeast quadrant in LTCs after landfall. This cyclonic rotation is found to be tightly related to a similar cyclonic rotation of environmental VWS vectors from South China to East China. The latter is determined by the large-scale circulations, including the western North Pacific monsoon trough, subtropical high, and the midlatitude westerly systems. This indicates that environmental VWS should be a useful predictor for the asymmetric rainfall distribution in LTCs. The effect of TC motion (including direction and speed) on the rainfall distribution in LTCs is secondary compared with the effect of environmental VWS. However, the land-sea contrast may predominantly control the asymmetric rainfall distribution in LTCs when the ambient VWS is weak, as illustrated in a schematic **Figure 21**. When the environmental VWS is weak (less than 5 m s^{-1}), the asymmetric rainfall maximum in the periphery of a LTC may shift from the downshear and offshore side (shown by the shaded area in **Figure 21a**) to the upshear and onshore side (shown by the shaded area in **Figure 21b**).

Note that most of the findings discussed in this chapter are based on the composite analyses of satellite-retrieved products, individual LTC may be considerably deviated from the composite because of other possible involved complex scale multiple interactions and the complicated effect of topography. Typhoons Utor (2013) and Fitow (2013) are shown as examples to illustrate how important some processes other than VWS are in shaping the rainfall distribution in LTCs. The latest study by Bell et al. [69] shows that the microphysics of precipitation also plays an important role in TC extreme events. The cloud microphysics can be strongly modulated by the flow over terrain, resulting in the changes of the dominant processes in upslope or downslope winds [70], as well as changes in precipitation efficiency [71]. Mesoscale terrain in the coastal region can affect the spatial distribution and duration of rainfall by deflecting the TC track [72].

Current studies have not been comprehensive yet in several topics related to rainfall distribution and evolution in LTCs. For example, the rainfall diurnal cycle in LTC has recently been revealed by Hu et al. [73], but more work is necessary to understand the involved physical

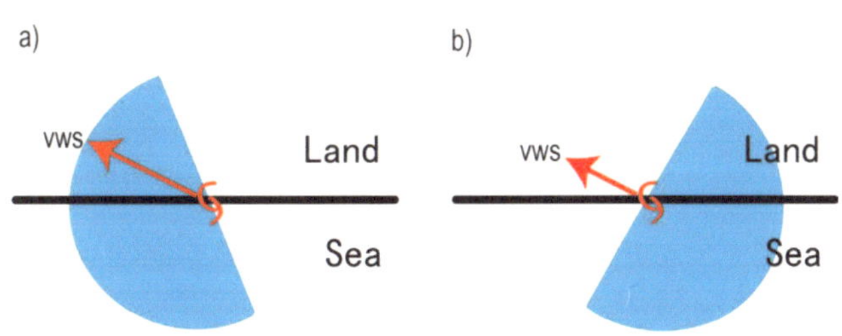

Figure 21. Schematic illustration of effects of both (a) strong and (b) weak environmental VWS (shown by the red arrows) combined with coastlines (shown by the bold black lines) on the rainfall asymmetries in LTCs.

mechanisms. The extreme rainfall in spiral rainbands is another topic not discussed/included here, so it needs to be thoroughly investigated because many extreme rainfall events in LTCs could be induced in spiral rainbands but not in the inner core region [74]. In addition, even a TC over the open ocean may also interact with neighboring weather systems, leading to rainfall in distant regions (over thousand kilometers from the TC), which is often called the TC remote effect [75]. These need to be comprehended further in future studies.

Finally, this chapter mainly summarizes the observational results and the related physical processes in determining both the axisymmetric and asymmetric distributions of rainfall in LTCs. Some recent efforts on improving the numerical prediction of TC-related rainfall have not been included. In fact, some recent studies have shown that radar data assimilation could substantially improve the extreme rainfall prediction skill, in particular for 0–24-h forecasts [76]. Nevertheless, this chapter provides a reference background for future studies on rainfall distribution in LTCs and for future efforts to improve rainfall forecast of LTCs not only in China but also in other regions of the world.

Acknowledgements

This work is supported in part by the National Basic Research and Development Project (973 program) of China under contract No. 2015CB452805 and in part by Key Program for International S&T Cooperation Projects of China (No. 2017YFE0107700).

Author details

Zifeng Yu[1] and Yuqing Wang[2,3]*

*Address all correspondence to: yuzf@typhoon.org.cn

1 Shanghai Typhoon Institute, China Meteorological Administration, Shanghai, China

2 State Key Laboratory of Severe Weather, Chinese Academy of Meteorological Sciences, China Administration, Beijing, China

3 Department of Atmospheric Sciences and International Pacific Research Center, University of Hawaii at Manoa, Honolulu, HI, USA

References

[1] Wang Y, Wu C-C. Current understanding of tropical cyclone structure and intensity changes – A review. Meteorology and Atmospheric Physics. 2004;**87**:257-278. DOI: 10.1007/s00703-003-0055-6

[2] Wang Y. Recent research progress on tropical cyclone structure and intensity. Tropical Cyclone Research and Review. 2012;**1**:254-275. DOI: 10.6057/2012TCRR02.05

[3] Marks Jr FD. Evolution of the structure of precipitation in Hurricane Allen (1980). Monthly Weather Review. 1985;**113**:909-930. DOI: 10.1175/1520-0493(1985)113<0909:EOTSOP>2.0.CO;2

[4] Merrill RT. Environmental influences on hurricane intensification. Journal of the Atmospheric Sciences. 1988;**45**:1678-1687. DOI: 10.1175/1520-0469(1988)045<1678:EIOHI>2.0.CO;2

[5] Jones SC. The evolution of vortices in vertical shear. I: Initially barotropic vortices. Quarterly Journal of the Royal Meteorological Society. 1995;**121**:821-851. DOI: 10.1002/qj.49712152406

[6] Jones SC. The evolution of vortices in vertical shear. II: Large-scale asymmetries. Quarterly Journal of the Royal Meteorological Society. 2000;**126**:3137-3160. DOI: 10.1002/qj.49712657008

[7] Jones SC. The evolution of vortices in vertical shear. III: Baroclinic vortices. Quarterly Journal of the Royal Meteorological Society. 2000;**126**:3161-3185. DOI: 10.1002/qj.49712657009

[8] Jones SC. On the ability of dry tropical-cyclone-like vortices to withstand vertical shear. Journal of the Atmospheric Sciences. 2004;**61**:114-119. DOI: 10.1175/1520-0469(2004)061<0114:OTAODT>2.0.CO;2

[9] Wang Y, Holland GJ. Tropical cyclone motion and evolution in vertical shear. Journal of the Atmospheric Sciences. 1996;**53**:3313-3332. DOI: 10.1175/1520-0469(1996)053<3313:TCMAEI>2.0.CO;2

[10] DeMaria M. The effect of vertical wind shear on tropical cyclone intensity change. Journal of the Atmospheric Sciences. 1996;**53**:2076-2088. DOI: 10.1175/1520-0469(1996)053<2076:TEOVSO>2.0.CO;2

[11] Frank WM, Ritchie EA. Effects of environmental flow upon tropical cyclone structure. Monthly Weather Review. 1999;**127**:2044-2061. DOI: 10.1175/1520-0493(1999)127<2044:EOEFUT>2.0.CO;2

[12] Frank WM, Ritchie EA. Effects of vertical wind shear on the intensity and structure of numerically simulated hurricanes. Monthly Weather Review. 2001;**129**:2249-2269. DOI: 10.1175/1520-0493(2001)129<2249:EOVWSO>2.0.CO;2

[13] Peng MS, Williams RT. Dynamics of vortex asymmetries and their influence on vortex motion on a β-plane. Journal of the Atmospheric Sciences. 1990;**47**:1987-2003. DOI: 10.1175/1520-0469(1990)047<1987:DOVAAT>2.0.CO;2

[14] Wang Y, Holland GJ. The beta drift of baroclinic vortices. Part I: Adiabatic vortices. Journal of the Atmospheric Sciences. 1996;**53**:411-427. DOI: 10.1175/1520-0469(1996)053<0411:TBDOBV>2.0.CO;2

[15] Wang Y, Holland GJ. The beta drift of baroclinic vortices. Part II: Diabatic vortices. Journal of the Atmospheric Sciences. 1996;**53**:3737-3756. DOI: 10.1175/1520-0469(1996)053<3737:TBDOBV>2.0.CO;2

[16] Bender MA. The effect of relative flow on the asymmetric structure in the interior of hurricanes. Journal of the Atmospheric Sciences. 1997;**54**:703-724. DOI: 10.1175/1520-0469(1997)054<0703:TEORFO>2.0.CO;2

[17] Peng MS, Jeng B-F, Williams RT. A numerical study on tropical cyclone intensification. Part I: Beta effect and mean flow effect. Journal of the Atmospheric Sciences. 1999;**56**:1404-1423. DOI: 10.1175/1520-0469(1999)056<1404:ANSOTC>2.0.CO;2

[18] Shapiro LJ. The asymmetric boundary layer flow under a translating hurricane. Journal of the Atmospheric Sciences. 1983;**40**:1984-1998. DOI: 10.1175/1520-0469(1983)040<1984:TABLFU>2.0.CO;2

[19] Dunion JP, Velden CS. The impact of the Saharan air layer on Atlantic tropical cyclone activity. Bulletin of the American Meteorological Society. 2004;**85**:353-365. DOI: 10.1175/BAMS-85-3-353

[20] Chen Y-S, Yau MY. Spiral bands in a simulated hurricane. Part I: Vortex Rossby wave verification. Journal of the Atmospheric Sciences. 2001;**58**:2128-2145. DOI: 10.1175/1520-0469(2001)058<2128:SBIASH>2.0.CO;2

[21] Wang Y. Vortex Rossby waves in a numerically simulated tropical cyclone. Part I: Overall structure, potential vorticity, and kinetic energy budgets. Journal of the Atmospheric Sciences. 2002;**59**:1213-1238. DOI: 10.1175/1520-0469(2002)059<1213:VRWIAN>2.0.CO;2

[22] Wang Y. Vortex Rossby waves in a numerically simulated tropical cyclone. Part II: The role in tropical cyclone structure and intensity changes. Journal of the Atmospheric Sciences. 2002;**59**:1213-1238. DOI: 10.1175/1520-0469(2002)059<1239:VRWIAN>2.0.CO;2

[23] Lonfat M, Marks Jr FD, Chen SS. Precipitation distribution in tropical cyclones using the tropical rainfall measuring mission (TRMM) microwave imager: A global perspective. Monthly Weather Review. 2004;**132**:1645-1660. DOI: 10.1175/1520-0493(2004)132<1645:PDITCU>2.0.CO;2

[24] Chen S, Knaff JA, Marks FD. Effects of vertical wind shear and storm motion on tropical cyclone rainfall asymmetries deduced from TRMM. Monthly Weather Review. 2006;**134**:3190-3208. DOI: 10.1175/MWR3245.1

[25] Yu Z, Wang Y, Xu H. Observed rainfall asymmetry in tropical cyclones making landfall over China. Journal of Applied Meteorology and Climatology. 2015;**54**:117-136. DOI: 10.1175/JAMC-D-13-0359.1

[26] Yu Z, Wang Y, Xu H, Davidson NE, Chen Y, Chen Y, Yu H. On the relationship between intensity and rainfall distribution in tropical cyclones making landfall over China. Journal of Applied Meteorology and Climatology. 2017;**56**:2883-2901. DOI: 10.1175/JAMC-D-16-0334.1

[27] Rogers R, Chen SS, Tenerelli J, Willoughby H. A numerical study of the impact of vertical shear on the distribution of rainfall in Hurricane Bonnie (1998). Monthly Weather Review. 2003;**131**:1577-1599. DOI: 10.1175//2546.1

[28] Lonfat M, Rogers R, Marchok T, Marks FD Jr. A parametric model for predicting hurricane rainfall. Monthly Weather Review. 2007;**135**:3086-3097. DOI: 10.1175/MWR3433.1

[29] Ueno M. Observational analysis and numerical evaluation of the effects of vertical wind shear on the rainfall asymmetry in the typhoon inner-core region. Journal of the Meteorological Society of Japan. 2007;**85**:115-136. DOI: 10.2151/jmsj.85.115

[30] Wingo MT, Cecil DJ. Effects of vertical wind shear on tropical cyclone precipitation. Monthly Weather Review. 2010;**138**:645-662. DOI: 10.1175/2009MWR2921.1

[31] Hence DA, Houze RA. Vertical structure of hurricane eyewalls as seen by the TRMM precipitation radar. Journal of the Atmospheric Sciences. 2011;**68**:1637-1652. DOI: 10.1175/2011JAS3578.1

[32] Reasor PD, Rogers R, Lorsolo S. Environmental flow impacts on tropical cyclone structure diagnosed from airborne Doppler radar composites. Monthly Weather Review. 2013;**141**:2949-2969. DOI: 10.1175/MWR-D-12-00334.1

[33] Tuleya RE, Kurihara Y. A numerical simulation of the landfall of tropical cyclones. Journal of the Atmospheric Sciences. 1978;**35**:242-257

[34] Tuleya RE, Bender MA, Kurihara Y. A simulation study of the landfall of tropical cyclones. Monthly Weather Review. 1984;**112**:124-136. DOI: 10.1175/1520-0493(1984)112<0124:ASSOTL>2.0.CO;2

[35] Jones RW. A simulation of hurricane landfall with a numerical model featuring latent heating by the resolvable scales. Monthly Weather Review. 1987;**115**:2279-2297. DOI: 10.1175/1520-0493(1987)115<2279:ASOHLW>2.0.CO;2

[36] Kepert JD, Wang Y. The dynamics of boundary layer jets within the tropical cyclone core. Part II: Nonlinear enhancement. Journal of the Atmospheric Sciences. 2001;**58**:2485-2501. DOI: 10.1175/1520-0469(2001)058<2485:TDOBLJ>2.0.CO;2

[37] Chan JCL, Liu KS, Ching SE, Lai EST. Asymmetric distribution of convection associated with tropical cyclones making landfall along the South China coast. Monthly Weather Review. 2004;**132**:2410-2420. DOI: 10.1175/1520-0493(2004)132<2410:ADOCAW>2.0.CO;2

[38] Kimball SK. Structure and evolution of rainfall in numerically simulated landfalling hurricanes. Monthly Weather Review. 2008;**136**:3822-3847. DOI: 10.1175/2008MWR2304.1

[39] Ramsay HA, Leslie LM, Kepert JD. A high-resolution simulation of asymmetries in severe southern hemisphere tropical cyclone Larry (2006). Monthly Weather Review. 2009;**137**:4171-4187. DOI: 10.1175/2009MWR2744.1

[40] Yu Z, Yu H, Gao S. Terrain impact on the precipitation of landfalling Typhoon Talim. Journal of Tropical Meteorology. 2010a;**16**:115-124. DOI: 10.3969/j.issn.1006-8775.2010.02.003

[41] Yu Z, Liang X, Yu H, Chan JCL. Mesoscale vortex generation and merging process: A case study associated with a post-landfall tropical depression. Advances in Atmospheric Sciences. 2010;**27**:356-370. DOI: 10.1007/s00376-009-8091-x

[42] Yu Z, Yu H. Application of generalized convective vorticity vector in a rainfall process caused by a landfalling tropical depression. Journal of Tropical Meteorology. 2012;**18**: 422-435

[43] Li Y, Cheung KKW, Chan JCL. Numerical study on the development of asymmetric convection and vertical wind shear during tropical cyclone landfall. Quarterly Journal of the Royal Meteorological Society. 2013;**140**:1866-1877. DOI: 10.1002/qj.2259

[44] Li Y, Cheung KKW, Chan JCL. Modelling the effects of land–sea contrast on tropical cyclone precipitation under environmental vertical wind shear. Quarterly Journal of the Royal Meteorological Society. 2014;**141**:396-412. DOI: 10.1002/qj.2359

[45] Xu W, Jiang H, Kang X. Rainfall asymmetries of tropical cyclones prior to, during, and after making landfall in South China and Southeast United States. Atmospheric Research. 2014;**139**:18-26. DOI: 10.1016/j.atmosres.2013.12.015

[46] Meng W, Wang Y. A diagnostic study on heavy rainfall induced by Typhoon Utor (2013) in South China: 1. Rainfall asymmetry at landfall. Journal of Geophysical Research-Atmospheres. 2016;**121**:12781-12802. DOI: 10.1002/2015JD024646

[47] Meng W, Wang Y. A diagnostic study on heavy rainfall induced by Typhoon Utor (2013) in South China: 2. Postlandfall rainfall. Journal of Geophysical Research-Atmospheres. 2016;**121**:12803-12819. DOI: 10.1002/2015JD024647

[48] Kepert JD. Observed boundary layer wind structure and balance in the hurricane core. Part I: Hurricane Georges. Journal of the Atmospheric Sciences. 2006;**63**:2169-2193.DOI: 10.1175/JAS3745.1

[49] Kepert JD. Observed boundary layer wind structure and balance in the hurricane core. Part II: Hurricane Mitch. Journal of the Atmospheric Sciences. 2006;**63**:2194-2211. DOI: 10.1175/JAS3746.1

[50] Chan JCL, Liang X. Convective asymmetries associated with tropical cyclone landfall. Part I: F-plane simulations. Journal of the Atmospheric Sciences. 2003;**60**:1560-1567. DOI: 10.1175/1520-0469(2003)60<1560:CAAWTC>2.0.CO;2

[51] Wong MLM, Chan JCL. Tropical cyclone motion in response to land surface friction. Journal of the Atmospheric Sciences. 2006;**63**:1324-1337. DOI: 10.1175/JAS3683.1

[52] Wong MLM, Chan JCL. Modeling the effects of land–sea roughness contrast on tropical cyclone winds. Journal of the Atmospheric Sciences. 2007;**64**:3249-3264. DOI: 10.1175/JAS4027.1

[53] Harnos SD, Nesbitt SW. Convective structure in rapidly intensifying tropical cyclones as depicted by passive microwave measurements. Geophysical Research Letters. 2011; **38**:L07805. DOI: 10.1029/2011GL047010

[54] Kieper EM, Jiang H. Predicting tropical cyclone rapid intensification using the 37 GHz ring pattern identified from passive microwave measurements. Geophysical Research Letters. 2012;**39**:L13804. DOI: 10.1029/2012GL052115

[55] Jiang H, Ramirez EM. Necessary conditions for tropical cyclone rapid intensification as derived from 11 years of TRMM data. Journal of Climate. 2013;**26**:6459-6470. DOI: 10.1175/JCLI-D-12-00432.1

[56] Alvey III GR, Zawislak J, Zipser E. Precipitation properties observed during tropical cyclone intensity change. Monthly Weather Review. 2015;**143**:4476-4492. DOI: 10.1175/MWR-D-15-0065.1

[57] Harnos SD, Nesbitt SW. Varied pathways for simulated tropical cyclone rapid intensification. Part I: Precipitation and environment. Quarterly Journal of the Royal Meteorological Society. 2016;**142**:1816-1831. DOI: 10.1002/qj.2780

[58] Jiang H, Halverson JB, Simpson J. On the differences in storm rainfall from hurricanes Isidore and Lili. Part I: Satellite observations and rain potential. Weather and Forecasting. 2008;**23**:29-43. DOI: 10.1175/2007WAF2005096.1

[59] Jiang H, Halverson JB, Zipser EJ. Influence of environmental moisture on TRMM-derived tropical cyclone precipitation over land and ocean. Geophysical Research Letters. 2008;**35**:L17806. DOI: 10.1029/2008GL034658

[60] Yu Z, Yu H, Chen P, Qian C, Yue C. Verification of tropical cyclone-related satellite precipitation estimates in Mainland China. Journal of Applied Meteorology and Climatology. 2009;**48**:2227-2241. DOI: 10.1175/2009JAMC2143.1

[61] Chen Y, Ebert EE, Walsh KJE, Davidson NE. Evaluation of TRMM 3B42 precipitation estimates of tropical cyclone rainfall using PACRAIN data. Journal of Geophysical Research – Atmospheres. 2013;**118**:1-13. DOI: 10.1002/jgrd.50250

[62] Chen Y, Ebert EE, Walsh KJE, Davidson NE. Evaluation of TMPA 3B42 daily precipitation estimates of tropical cyclone rainfall over Australia. Journal of Geophysical Research – Atmospheres. 2013;**118**:1-13. DOI: 10.1002/2013JD020319

[63] Kalnay E, et al. The NCEP/NCAR 40-year reanalysis project. Bulletin of the American Meteorological Society. 1996;**77**:437-471. DOI: 10.1175/1520-0477(1996)077<0437:TNYRP>2.0.CO;2

[64] Shu S, Ming J, Chi P. Large-scale characteristics and probability of rapidly intensifying tropical cyclones in the Western North Pacific Basin. Weather and Forecasting. 2012;**27**:411-423. DOI: 10.1175/WAF-D-11-00042.1

[65] Boyd JP. Chebyshev and Fourier Spectral Methods. 2nd ed. New York: Dover; 2001. 44 p

[66] Chen Y-S, Yau M Y. Asymmetric structures in a simulated landfalling hurricane. Journal of the Atmospheric Sciences. 2003;**60**:2294-2312. DOI: /10.1175/1520-0469(2003)060<2294:ASIASL>2.0.CO;2

[67] Bao X, Davidson NE, Yu H, Hankinson M, Sun Z, Rikus LJ, Liu J, Yu Z, Wu D. Diagnostics for an extreme rain event near Shanghai, during the landfall of Typhoon Fitow (2013). Monthly Weather Review. 2015;**143**:3377-3405. DOI: 10.1175/MWR-D-14-00241.1

[68] Yu Z, Chen Y, Ebert B, Davidson N E, Xiao Y, Yu H, Duan Y. Benchmark rainfall verification of ACCESS-TC operational landfall forecasts over China. Meteorological Applications. 2018 (submitted)

[69] Bell MM, Kingsmill DE, White AB. The ice water paths of small and large ice species in Hurricanes Arthur (2014) and Irene (2011). Journal of Applied Meteorology and Climatology. 2017;**56**:1383-1404. DOI: 10.1175/JAMC-D-16-0300.1

[70] DeHart JC, Houze RA. Orographic modification of precipitation processes in Hurricane Karl (2010). Monthly Weather Review. 2017;**145**:4171-4186. DOI: 10.1175/MWR-D-17-0014.1

[71] Huang H-L, Yang M-J, Sui C-H. Water budget and precipitation efficiency of Typhoon Morakot (2009). Journal of the Atmospheric Sciences. 2014;**71**:112-129. DOI: 10.1175/JAS-D-13-053.1

[72] Lin YL, Chen SH, Liu L. Orographic influence on basic flow and cyclone circulation and their impacts on track deflection of an idealized tropical cyclone. Journal of the Atmospheric Sciences. 2016;**73**:3951-3974. DOI: 10.1175/JAS-D-15-0252.1

[73] Hu H, Duan Y, Wang Y, Zhang X. Diurnal cycle of rainfall associated with landfalling tropical cyclones in China from rain-gauge observations. Journal of Applied Meteorology and Climatology. 2017;**56**:2595-2605. DOI: 10.1175/JAMC-D-16-0335.1

[74] Lin Y, Li Y, Li Q, Chen M, Xu F, Wang Y, Huang B. A long lasting vortex Rossby wave induced rainband of Typhoon Longwang (2005). Bulletin of the American Meteorological Society. 2018;**99** (in press). DOI: 10.1175/BAMS-D-17-0122.1

[75] Wang Y, Wang Y, Fudeyasu H. The role of Typhoon Songda (2004) in producing distantly located heavy rainfall in Japan. Monthly Weather Review. 2009;**137**:3699-3716. DOI: 10.1175/2009MWR2933.1

[76] Bao X, Wu D, Lei X, Ma L, Wang D, Zhao K, Jou BJ. Improving extreme rainfall forecast of Typhoon Morakot (2009) by assimilating radar data from Taiwan and mainland China. Journal of Meteorological Research. 2017;**31**:747-766. DOI: 10.1007/s13351-017-6007-8

Chapter 5

Extreme Weather Events in Ukraine: Occurrence and Changes

Vira Balabukh, Olena Lavrynenko,
Volodymyr Bilaniuk, Andriy Mykhnovych and
Olha Pylypovych

Additional information is available at the end of the chapter

http://dx.doi.org/10.5772/intechopen.77306

Abstract

Extreme weather is in the attention focus of many scientists and managers during the last decades. The leading aspect of these phenomena investigations in the recent years is the risk of material and human losses and damage mitigation. Especially, the interest is with regard to effects of weather extremities on natural systems and social processes such as land use practices, water resources management, emergency management, and planning. The main objectives of the investigations are clarifying of spectrum, space and time regularities of extreme weather events occurring in Ukraine as well as their intensity, duration, daily and seasonal variation, spreading, recurrence in the regions, and their changes analyzed. Applying statistical and geographical space–time analyses, the main regularities of the extreme weather events' occurrence have been described as well as the trends and intensity of the extreme weather regime changes in Ukraine have been calculated and assessed.

Keywords: Ukraine, extreme weather event, recurrence, changes trends

1. Introduction

Weather event can be defined as extreme when it reaches a certain boundary value. In Ukraine, about 20 weather phenomena can be called as the extreme weather event. They are observed by the State Hydrometeorology Committee. These phenomena occur on the whole territory of Ukraine with different recurrence and intensity (**Table 1**) [1].

Phenomenon	Events amount	%	Amount of points	%
Heavy rain	1335	44.0	3712	52.9
Heavy snowfall	237	7.8	462	6.6
Hail	131	4.3	287	4.1
Strong wind	398	13.1	989	14.0
Squall	164	5.4	234	3.3
Spout	16	0.5	47	0.7
Sand storm	3	0.1	33	0.5
Snowstorm	249	8.2	427	6.1
Fog	292	9.6	536	7.6
Glazed frost	68	2.3	139	2.0
Soggy snow adhering	68	2.3	79	1.1
Compound ice deposits	70	2.4	78	1.1
Total	3031	100	7023	100

Table 1. Extreme weather phenomena and their occurrence in Ukraine and the amount of observation points where the phenomenon has been observed in 1986–2010 [1].

The duration, spreading, recurrence, and intensity of extreme weather events will increase during this century both globally and regionally [2], and many practical problems require relevant information on extreme event changes. According to the Fifth Assessment Report of the Intergovernmental Panel on Climate Change [3], the main effects of the regional climate changes due to the global warming are not only a significant increase in air temperature, in the heat waves, changes of thermal regime [4, 5], precipitation structure, and an increase in the number of heavy precipitation [4], but also an increase in the frequency and intensity of extreme weather events [2, 6, 7] and their effects like alteration of ecosystems, disruption of food production and water supply, damage to infrastructure and settlements, human mortality, and negative consequences for human well-being [3]. Losses from weather-related disasters have increased substantially in recent decades.

The scientists are still debating the best methods to account climate change in extreme weather value analysis. The wide range of available and applicable methods is presented in [2]. In Ukraine, the main definitions of the extreme weather events as well as the methods and instructions on their analysis and prediction are presented in [8, 9].

The trends, scales, and intensity of climate and weather extremes changes are analyzed in global [4, 6], regional [10], and national [1, 7, 11] aspects. There are a few papers on the occurrence and spreading of the main weather extreme phenomena on different time periods [11–14]. But since extreme weather is very unstable and changeable in space and time, it is necessary to complete and actualize existing knowledge base.

The climatologic information has been presented in our research according to administrative division by the regions, groups of the regions, and for Ukraine as a whole (**Figure 1**).

http://dx.doi.org/10.5772/intechopen.77306

Figure 1. Study area.

To analyze the spectrum of the regularly observed extremely dangerous meteorological phenomena and to evaluate the long-term changes of their recurrence in Ukraine, two periods have been worked out. As the base, the period 1961 (1971)–1990 has been chosen, and as the recent one the period 1981–2010 has been worked out. The linear trends of the extreme weather events amount changes as well as tendencies of their intensity changes have been analyzed. The statistical significance of the linear trends has been assessed as well. Most of events recurrence was analyzed for a point of observations. Since extreme squalls, hails are characterized by poor recurrence in the observation point and spread on local areas; therefore, to evaluate the risk of these phenomena development and their space–time changes, the events amount was adjusted to 100 km^2 area (β-mesoscale).

2. Extreme weather events occurrence, characteristics, and changes

2.1. Heavy rain

In Ukraine, the rain becomes a dangerous weather event when it reaches 15–50 mm per 12 h (15–30 mm per 12 h for mud flow and erosion threat areas) and distress event when exceed 50 mm per 12 h (30 mm per 12 h for mud flow and erosion threat areas). Other extreme weather events are heavy rainfalls (>30 mm per hour) and long-continued rain (duration more than 1 day and precipitation is 100 mm and more) [8].

The recurrence of all types of heavy rains has a good marked seasonal course. All rainfalls, 94% of heavy rains and 88% of long-continued rains, are observed in the warm period. The maximum of the extremes is observed in July. Rainfalls dominate and long-continued rains are seldom. The share of extreme rains in winter period is 6–12%. Long-continued rains dominate in winter, especially in the western region.

As usual, extreme rains start afternoon (most of distress rains start between 12 and 18 o'clock by maximal convection). Long-continued rains start in the night hours.

During distress rain (>50 mm per 12 h and less), most often (77%), precipitation sum is 50–70 mm. Rains with 100 mm are very rare (3.5%) and locate in the southern and central regions of Ukraine as usual. In the western region, extreme rains have the highest recurrence. The maximum of precipitation per 12 h in Ukraine has been fixed 9.05.1968 in Voznesens'k City in Mykolaiv region (151 mm). Mostly, the extreme rainfalls (86%) have intensity of 30–50 mm per hour and less. Sometime (probability <1%), rain intensity can reach 100 mm per hour. The intensity and recurrence of the rainfalls are highest in the South of Ukraine. The maximum (122 mm per hour) was observed 9.05.1968 in Voznesens'k City.

Most of long-continued rains (83%) are characterized by 100–150 mm precipitation sum. A share of rains with more than 200 mm sum in Ukraine is about 1%. In the South, it can reach 5%. The longest rains are observed in Western and Southern Ukraine. The absolute maximum was observed in Crimean Mountains (Ay-Petri) 5–6.09.1968 (207 mm).

In Ukraine, heavy rains and rainfalls are observed annually, long-continued rains—two times per 3 years. The rains of 50 mm and more per 12 h and less are observed annually in all regions. Heavy rainfalls are observed every year in the southern, western, and central regions. Long-continued rains are observed every 3–4 years in the southern and western regions. In central and northern regions, they are observed every 6–8 years and are very rare in the eastern region. Heavy rainfalls are very local (on the territory of one or two administrative districts), but heavy and long-continued rains can occupy a few administrative regions especially in the West and South.

The recurrence of extreme rains is very changeable in space and time, but it decreases from the North and North West to South and South East. The highest recurrence (about two events in a station) is in Ivano-Frankivs'k and Transcarpathian regions in mountains (extreme rain reaches about 40 mm per 12 h in average and 100 mm and more as maximum). Also, a high recurrence (5–8 per 10 years in a station) is observed in Lviv, Ternopil, Chernivtsi regions, and Crimea, especially in the mountains and highlands (intensity is from 50 mm (average) to >150 mm (maximum) per 12 h). Rarely (less than one event per 10 years), extreme rains are observed in the North and North East (from 60 to 70 mm (average) to 100–130 mm (maximum) per 12 h). The recurrence of extreme rains depends also on the altitude. Events amount, duration, and intensity increase with altitude. At that, the amount of dangerous rains increases faster than distress ones. The altitude 500–700 m is most optimal for all types of extreme rain occurrence. The maximum of distress rains recurrence is observed in the Carpathians and Crimean Mountains. Another factor of recurrence is water surface. Above waters and coasts, the recurrence of extreme rains in spring and the start of summer is less than over land. In April and June, along the Black and Azov Sea shore, 25–35% of the extreme rains are observed. At the same time, at the distance of 100–150 km, this parameter is 35–50%.

The amount of extreme rain events in Ukraine changes every year. At that with the probability of 99% during the last 40 years (1971–2010), their recurrence increases. The amount of precipitation and its maximal sum also increase. These trends are characteristic almost for the whole territory of Ukraine but most marked trends are ascertained in the western, northern, and southern regions, especially in Ivano-Frankivs'k, Chernivtsi, Lviv, Rivne, Kherson, and Kropyvnyts'kyi administrative regions (**Table 2**, **Figure 2**). In the East and the Centre low increasing of the rains amount and decreasing of their intensity are observed. And only in Donets'k,

Region	Administrative region	Recurrence				Average intensity				Maximal intensity			
		Distress rains		Dangerous rains		Distress rains		Dangerous rains		Distress rains		Dangerous rains	
		α	p	α	p	α	p	α	p	α	p	α	p
West	Volynian	0.002	0.93	0.594	0.00	0.051	0.90	0.140	0.01	−0.0001	0.99	0.292	0.02
	Transcarpathian	0.149	0.26	0.136	0.43	−0.021	0.62	0.014	0.15	0.160	0.34	0.052	0.06
	Ivano-Frankivs'k	0.224	0.01	0.324	0.06	0.026	0.82	0.008	0.44	0.684	0.02	0.010	0.62
	Lviv	0.086	0.23	0.394	0.09	0.150	0.17	0.035	0.01	0.387	0.09	0.127	0.09
	Rivne	0.018	0.15	0.186	0.02	0.758	0.11	0.109	0.06	0.780	0.10	0.235	0.07
	Ternopil	0.008	0.75	−0.023	0.81	0.029	0.91	0.031	0.13	0.172	0.63	0.169	0.08
	Khmelnyts'kyi	0.018	0.48	0.126	0.33	−0.175	0.51	0.047	0.02	−0.058	0.86	0.173	0.03
	Chernivtsi	0.085	0.03	0.055	0.43	0.207	0.30	0.041	0.03	0.319	0.34	0.022	0.48
South	Zaporizhzhia	0.030	0.21	0.155	0.30	0.185	0.67	0.133	0.02	0.289	0.52	0.267	0.04
	Crimea	0.127	0.46	0.385	0.27	−0.107	0.09	0.087	0.08	0.226	0.65	0.191	0.10
	Mykolaiv	0.002	0.87	0.036	0.77	0.579	0.21	0.102	0.06	0.521	0.29	0.150	0.21
	Odesa	−0.003	0.87	0.335	0.02	0.339	0.31	0.009	0.63	0.600	0.18	0.052	0.42
	Kherson	0.033	0.27	0.225	0.16	1.098	0.02	0.109	0.04	1.344	0.01	0.244	0.04
North	Zhytomyr	0.009	0.48	0.251	0.05	0.032	0.95	0.085	0.10	0.032	0.95	0.126	0.28
	Kyiv	0.004	0.87	0.324	0.13	−0.570	0.14	0.021	0.35	−0.354	0.43	0.120	0.19
	Sumy	0.0008	0.95	0.010	0.94	0.014	0.98	0.095	0.06	0.016	0.97	0.124	0.28
	Chernihiv	0.010	0.48	0.206	0.18	0.015	0.98	−0.016	0.45	0.042	0.94	0.060	0.46
East	Donets'k	−0.012	0.74	0.067	0.64	−0.299	0.33	0.005	0.84	−0.437	0.23	0.092	0.40
	Luhans'k	0.0003	0.98	0.148	0.20	−0.044	0.93	−0.020	0.47	−0.075	0.88	0.107	0.27
	Kharkiv	0.020	0.41	0.079	0.70	0.376	0.41	0.011	0.63	0.513	0.33	0.122	0.15
Centre	Vinnytsia	−0.033	0.13	−0.158	0.33	0.067	0.89	0.008	0.71	−0.244	0.67	0.027	0.67
	Dnipro	−0.002	0.91	0.105	0.54	−0.413	0.42	0.116	0.03	−0.463	0.39	0.201	0.08
	Kropyvnyts'kyi	0.042	0.02	0.543	0.00	0.136	0.72	0.028	0.31	0.271	0.56	0.060	0.48
	Poltava	−0.002	0.89	−0.153	0.23	0.421	0.34	0.137	0.01	0.471	0.34	0.184	0.16
	Cherkasy	0.008	0.65	0.420	0.02	−0.545	0.20	0.007	0.78	−0.541	0.24	0.108	0.23
Ukraine		0.948	0.02	4.992	0.05	−0.051	0.21	0.023	0.00	0.366	0.17	0.040	0.06

Table 2. Linear trend coefficient (α) and its significance (p) of recurrence (amount of events) and the intensity (average and maximal) of the extreme rains in observation point in Ukraine and its regions.

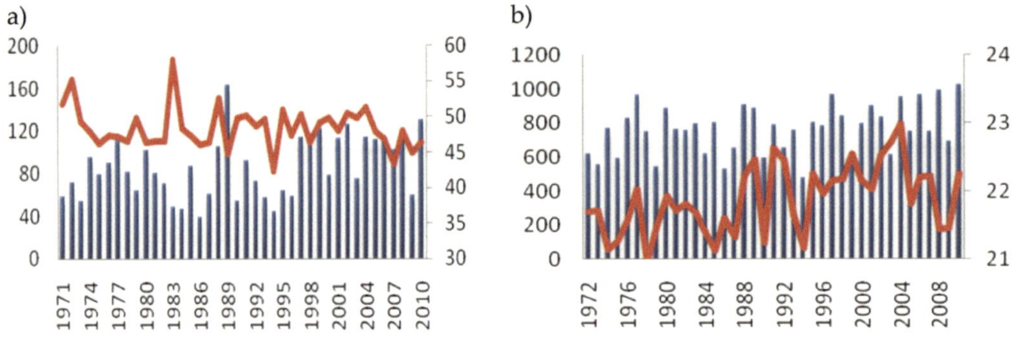

Figure 2. Changes of recurrence and intensity of the distress (a) and dangerous (b) rain events in Ukraine. Blue columns—amount of events. Red line—average precipitation, mm.

Dniptopetrovs'k and Vinnytsia regions, there is a weak tendency to decrease the recurrence and intensity. In the South and Center, the distress rains intensity has significantly increased.

The most noticeable changes of the recurrence and intensity of extreme rains in warm period are observed from the end of twentieth—the start of the twenty-first centuries (**Figure 2**). The changes are caused by an average troposphere temperature increase, humidity increase, and the intensification of convection and rain-forming processes [11]. Another significant cause of these changes is transformation of atmosphere circulation reflected in increasing of meridional processes in cyclones forming, changes of regions of cyclones forming, their moving trajectory and intensity.

2.2. Heavy snowfalls

In Ukraine, the snowfall becomes extreme weather event when precipitation sum reaches 20 mm per 12 h and less. Such snowfalls cause disturbance in the economy, transport, communications, and avalanches activization. The most dangerous is combining heavy snowfalls with freeze and strong wind.

In Ukraine, 85% of heavy snowfalls are characterized by an average precipitation of 20–30 mm per 12 h and are observed almost in all regions. The exclusion is mountain areas of the Carpathians and Crimea, Zaporizhzhia, Dnipro, Kyiv, and Sumy regions, where snowfalls are characterized by a higher intensity. The maximal precipitation on the plain territory of Ukraine can reach 40 mm, in the mountains—100 mm and more. The absolute maximum in the period of 1971–2010 has been observed 14–15.02.1997 on the Ay-Petri Mountain (132 mm).

Heavy snowfalls in Ukraine usually occur in the period October–April, but are possible in September and May as well (1966, 1989, 1991, 1992). Their recurrence has a good marked seasonal course. They are observed usually in winter (62%) and reach a maximum in January. In spring, about 22% and in autumn almost 14% of extreme snowfalls are observed. Heavy snowfalls are formed in night and morning hours as usual.

The average duration of extreme snowfalls varies from 9 to 12 h. The most long-continued snowfalls are observed in the West, the shortest duration in the South. The minimal duration is 1–2 h. The absolute minimum has been fixed 15.03.1998 on the Ay-Petri Mountain (0.9 h) with 21-mm precipitation. The longest duration snowfalls are observed in autumn and spring, the shortest duration in January.

In Ukraine, the extreme snowfalls are observed almost every year. Such a recurrence is typical for southern and western regions. In the North, they are observed once per 2 years, and in the East once per 4–5 years. Most often, heavy snowfalls in observation point are fixed in Ivano-Frankivs'k and Transcarpathian regions (two to four events per 10 years), a maximum in Carpathian (1–2 per decade) and Crimean Mountains. Noticeable recurrence (5–8 events per 100 years in meteostation) is observed also on the Volynian-Podolian Highland, in Ternopil, Khmelnyts'kyi, Zhytomyr, Vinnytsia, and Kropyvnyts'kyi regions. Less occurrence is characteristic for the Black Sea Lowland and Azov Highland, in Zaporizhzhia, Mykolaiv, Donets'k, and Dnipro regions. The least recurrence is observed on western and southern slopes of the Middle-Rus' Highland, in Sumy, Kharkiv, Luhans'k regions (two events per 100 years in meteostation).

Usually, the highest intensity of extreme snowfalls is characteristic for those regions which are characterized by a highest recurrence. In the Ukrainian Carpathians, during extreme snowfalls, the intensity is about 26–30 mm per 12 h in average and more than 100 mm per 12 h maximum. In the Crimean Mountains, the average intensity is 22–23 mm and maximal can reach 80 and more per 12 h. In the Black Sea Lowland and Azov Upland, in Zaporizhzhia, Mykolaiv, Donets'k, Dnipro, and Odesa regions, the average intensity of snowfalls is 24–25 mm and the maximal is 38 mm per 12 h. Less intensity is observed on the western and southern slopes of the Middle-Rus' Highland, in Sumy, Kharkiv, and Luhans'k regions, where heavy snowfalls occupy smaller areas. The least intensity is observed in the North-Western part of Ukraine, in Polissia.

In 1970s, a stable decrease of extreme snowfalls amount has been observed. The minimum has been reached in the middle of 1990s. Since the end of the twentieth century, recurrence and intensity started to increase and continue until now. The regional features of these changes are presented in **Table 3** and **Figure 3**.

With 99% of probability ($p \leq 0.01$), we can say about a significant increase of extreme snowfalls amount in Ukraine. The intensity of increase equals four events per 10 years in the observation point. At that, largest changes are observed in the South, East, and West, where recurrence and intensity (especially maximal one) during snowfall increased. The Biggest changes concern the Ukrainian Carpathians and Crimean Mountains.

In the central regions, the amount of snowfalls has not changed noticeably except Kropyvnyts'kyi and Cherkasy regions where the increase is 10–15%. Decreasing of recurrence is most noticeable in Lviv and Ternopil regions and a little less in the northern regions except Chernihiv. The intensity has been decreased significantly in the North, Lviv, and Ternopil regions as well during 1971–2010. These decreases reach 5–7 mm per decade with 99% probability (**Table 3**). During the last 40 years, the locality of extreme snowfalls is changed due to a decrease of recurrence and intensity within the Volynian-Podolian Highland and a noticeable increase of these parameters in Black Sea Lowland and Azov Highland.

2.3. Glazed frost

Glazed frost is a layer of suffused or transparent risen on the objects due to the freezing of rain drops frizzle or fog. The intensity of glazed frost is defined by the thickness of the ice layer, the velocity of the ice layer rising per time unit, and the duration of the phenomenon. In Ukraine, the glazed frost becomes a dangerous event when the ice layer thickness is

	Administrative region	Recurrence		Average intensity		Maximal intensity	
		α	p	α	p	α	p
West	Volynian	0.004	0.542	0.043	0.521	0.046	0.516
	Transcarpathian	0.082	0.077	0.232	0.171	0.368	0.120
	Ivano-Frankivs'k	0.095	0.001	0.392	0.037	0.635	0.006
	Lviv	−0.030	0.049	−0.195	0.177	−0.261	0.135
	Rivne	0.002	0.626	0.048	0.589	0.048	0.589
	Ternopil	−0.014	0.069	−0.204	0.136	−0.211	0.129
	Khmelnyts'kyi	−0.014	0.150	−0.111	0.445	−0.123	0.405
	Chernivtsi	−0.002	0.733	−0.021	0.844	−0.021	0.844
South	Zaporizhzhia	0.019	0.179	0.210	0.150	0.224	0.175
	Crimea	0.112	0.027	0.184	0.361	0.258	0.486
	Mykolaiv	0.008	0.315	0.313	0.041	0.308	0.047
	Odesa	0.022	0.224	0.295	0.074	0.306	0.077
	Kherson	0.005	0.411	0.126	0.214	0.126	0.218
North	Zhytomyr	−0.009	0.266	−0.187	0.146	−0.185	0.154
	Kyiv	−0.008	0.501	−0.183	0.179	−0.208	0.163
	Sumy	−0.005	0.522	−0.100	0.469	−0.099	0.493
	Chernihiv	0.010	0.428	−0.017	0.912	−0.010	0.951
East	Donets'k	0.026	0.069	0.164	0.182	0.214	0.125
	Luhans'k	0.016	0.160	0.228	0.032	0.234	0.041
	Kharkiv	0.014	0.295	0.197	0.153	0.206	0.153
Centre	Vinnytsia	−0.009	0.544	0.067	0.625	0.042	0.779
	Dnipro	0.022	0.116	0.174	0.258	0.199	0.223
	Kropyvnyts'kyi	0.026	0.095	0.264	0.049	0.281	0.045
	Poltava	−0.021	0.117	0.012	0.930	−0.015	0.913
	Cherkasy	0.007	0.575	0.227	0.122	0.221	0.143
Ukraine		0.389	0.012	0.051	0.560	0.116	0.696

Table 3. Linear trend coefficient (α) and its significance (p) of recurrence (amount of events) and the intensity (average and maximal) of the extreme snowfalls in observation point in Ukraine and its regions in 1971–2010.

6–19 mm and distress when it exceeds 20 mm [2]. Such events cause damage and losses in the energy, transport, communications, domestic economy, and so on. Glazed frost with a thickness of more than 20 mm in Ukraine is observed from October to April with two maximums: November and February–March. Extreme-gazed frost occurs in morning (3–6) and evening hours (16–20) [8].

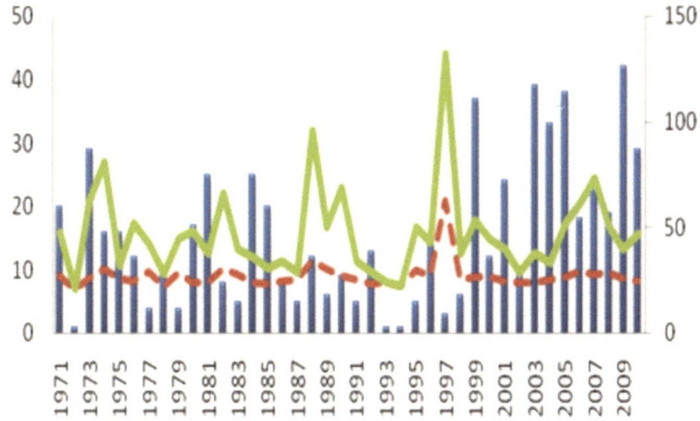

Figure 3. Changes of recurrence and intensity of the heavy snowfalls events in Ukraine. Blue columns—amount of events. Red line—average precipitation, mm. Green line—maximal precipitation, mm.

The average thickness of dangerous glazed frost ice layer varies between 9 and 17 mm, of the distress one—between 20 and 60 mm. This thickness dominates almost in all regions except north-western regions where it does not exceed 10 and 40 mm accordingly. Ice sediments of more than 100 mm are observed very rare, advantageously in the Crimea, and amount 1% of events. But there are events with the ice sediment thickness more than 200 mm. In November–December 2000, the ice sediments of 207 mm have been fixed in Zatyshshia village (Odesa region) and 307 mm in Karabi-Yayla (Crimea) in March 1977. Distress glazed frost is characterized by the high intensity. In 50% of the events, it amounts 1.1–2.0 mm per hour. In the Ay-Petri (Crimea), it reaches 5 mm per hour and more [11].

The overall duration of glazed frost in Ukraine fluctuates between a few hours and a few days and amounts about 60 h in average. The maximal duration can reach 100–200 h and even more. Usually, it is observed in the central and south-western regions. Most often (54%), ice deposits of more than 20 mm maintain less than 24 h. About 70% of the glazed frost events continue up to 2 days. Very rarely, extreme glazed frost can continue more than 1 week. For example, in November 2000, in Zhmerynka city (Vinnytsia region), the glazed frost has continued 250 h, in Vinnytsia city 282 h. The most long-continued extreme glazed frost maintains in November and February in the East and South-East.

The spatial occurrence of the extreme glazed frost is characterized by significant inequality. The highest recurrence (5–9 events per 10 years in observation point) of dangerous glazed frost with ice thickness more than 6 mm is observed in the East and South-East of Ukraine, especially within the Donets'k Range and the Azov Highland. The distress glazed frost with more than 20-mm ice layer in these regions is observed once per 3–5 years. The average ice layer thickness amounts to 10–12 mm, the maximum can reach 70 mm and more. A significant recurrence of extreme glazed frost (3–8 events per decade) is observed in the central part of the Black Sea Lowland and in the eastern part of the Dnipro Highland. These regions are characterized also by a highest intensity of extreme glazed frost. The average thickness changes between 10 and 17 mm, the maximal one can reach 200 mm and more. Such parameters are caused by the main course of the southern and south-western cyclones moving through these

	Administrative region	Recurrence		Average		Maximal	
		α	p	α	p	α	p
West	Volynian	0.02	0.09	0.07	0.31	0.09	0.23
	Transcarpathian	0.05	0.10	0.27	0.10	0.27	0.51
	Ivano-Frankivs'k	−0.02	0.42	−0.08	0.34	−0.18	0.13
	Lviv	−0.06	0.01	−0.12	0.07	−0.16	0.06
	Rivne	−0.02	0.08	−0.05	0.31	−0.11	0.15
	Ternopil	−0.06	0.002	−0.11	0.12	−0.15	0.14
	Khmelnyts'kyi	−0.06	0.01	−0.05	0.55	−0.04	0.76
	Chernivtsi	−0.004	0.73	−0.04	0.58	−0.05	0.56
South	Zaporizhzhia	−0.03	0.33	−0.02	0.81	−0.05	0.65
	Crimea	−0.11	0.05	−0.30	0.29	−1.17	0.08
	Mykolaiv	−0.004	0.89	−0.12	0.37	−0.15	0.47
	Odesa	−0.11	0.03	0.15	0.42	0.24	0.60
	Kherson	−0.11	0.20	−0.13	0.12	−0.25	0.23
North	Zhytomyr	0.01	0.55	0.01	0.81	0.06	0.44
	Kyiv	−0.002	0.93	−0.06	0.39	−0.09	0.34
	Sumy	−0.002	0.89	−0.12	0.07	−0.11	0.17
	Chernihiv	−0.02	0.26	−0.02	0.82	0.01	0.93
East	Donets'k	−0.12	0.13	0.03	0.84	−0.003	0.98
	Luhans'k	−0.02	0.59	−0.13	0.14	−0.41	0.07
	Kharkiv	0.01	0.77	−0.01	0.88	−0.02	0.93
Centre	Vinnytsia	−0.07	0.07	−0.06	0.42	−0.10	0.45
	Dnipro	−0.002	0.96	−0.04	0.56	0.01	0.93
	Kropyvnyts'kyi	−0.12	0.06	−0.10	0.17	−0.17	0.24
	Poltava	−0.12	0.001	−0.19	0.003	−0.30	0.01
	Cherkasy	−0.02	0.53	−0.04	0.58	−0.08	0.38
Ukraine		−0.67	0.14	0.08	0.12	−0.13	0.80

Table 4. Linear trend coefficient (α) and its significance (p) of recurrence (amount of events) and the ice deposit thickness (average and maximal) of the extreme glazed frost in observation point in Ukraine and its regions in 1971–2010.

regions and the strong frontal glazed frost forming with high intensity. Distress glazed frost (more than 20 mm) in these regions is observed every 5–10 years and can occupy more than two administrative districts. A much less recurrence (2–3 times per decade in meteostation) is characteristic for the slopes of Volynian-Podolian Highland and Ukrainian Carpathians. The average ice layer amounts to 9–11 mm, the maximal value can reach 60 mm and more.

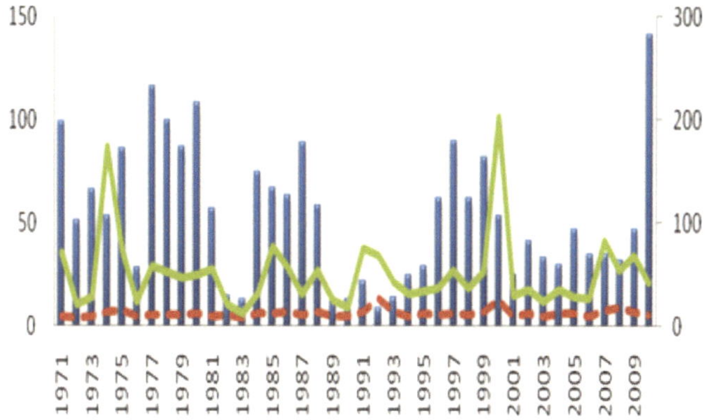

Figure 4. Changes of recurrence and ice deposit thickness of the extreme glazed frost events in Ukraine. Blue columns—amount of events. Red line—average thickness, mm. Green line—maximal thickness, mm.

The least recurrence (1–2 times per decade in the point) of extreme glazed frost is observed in Polissia Lowland. Here, the phenomena are with the least intensity and the average ice layer between 8 and 10 mm and the maximal value of 23–34 mm.

Since extreme glazed frost belongs to weather phenomena which are characterized by least recurrence, to evaluate the long-term changes, both kinds of glazed frost (dangerous and distress) were analyzed. With the 85% probability, we can say about a decrease in extreme glazed frost recurrence since 1971 in the whole territory of Ukraine. The most significant decrease in recurrence and ice layer thickness is observed in the West (**Table 4**, **Figure 4**), especially noticeably in Lviv region (probability >95%). Volynian and Transcarpathian regions are excluded, where noticeable increase in both parameters is observed (**Table 4**). A noticeable decrease of the analyzed parameters is the characteristic for the central and southern regions as well. The most significant changes in these regions are observed in Poltava region and Crimea. A feeble decrease in recurrence and ice layer thickness is observed also in the northern and eastern regions (**Table 4**).

2.4. Soggy snow adhering

In Ukraine, the soggy snow adhering becomes a dangerous weather event when the thickness of deposits amounts to 11–34 mm and distress one when the thickness exceeds 35 mm. At that only in 6% of soggy snowfalls, the adhering process is observed [8]. The main parameter of this phenomenon is its diameter in mm. About 80% of the events amount to 35–60 mm, about 2% of events have a diameter of >100 mm. The most intensive adhering is observed in the western and southern regions, where events of 80 mm amount to >5%, in the northern regions 3%. The level of danger depends on the phenomenon duration. The average duration in Ukraine fluctuates from 14 to 19 h. The maximum varies from 2 days to a week and more. The longest soggy snow adhering is observed in the West of Ukraine. But absolute maximum was fixed in Olevs'k City (Zhytomyr region) 12.02.1983 (304 h).

Extreme soggy snow adhering with distress diameter occurs in October–April usually. Most often (about 70%), it is in December–February. About 20% of events arise in evening hours, from 18 to 21 o'clock.

	Administrative region	Recurrence		Average diameter		Maximal diameter	
		α	p	α	p	α	p
West	Volynian	0.03	0.14	0.13	0.49	0.16	0.47
	Transcarpathian	0.06	0.09	−0.01	0.97	0.75	0.15
	Ivano-Frankivs'k	−0.04	0.01	−0.46	0.05	−1.07	0.007
	Lviv	−0.03	0.19	−0.08	0.67	−0.22	0.39
	Rivne	−0.01	0.55	−0.22	0.20	−0.23	0.25
	Ternopil	−0.003	0.82	−0.04	0.84	−0.05	0.82
	Khmelnyts'kyi	−0.002	0.94	−0.10	0.55	−0.10	0.69
	Chernivtsi	0.01	0.05	0.24	0.09	0.24	0.09
South	Zaporizhzhia	0.01	0.47	−0.22	0.30	−0.17	0.51
	Crimea	0.06	0.02	0.26	0.24	0.57	0.15
	Mykolaiv	−0.04	0.02	−0.26	0.15	−0.42	0.06
	Odesa	0.04	0.13	0.34	0.02	0.38	0.05
	Kherson	0.02	0.11	0.18	0.36	0.23	0.30
North	Zhytomyr	0.02	0.49	−0.05	0.80	−0.13	0.69
	Kyiv	0.06	0.08	0.03	0.88	0.10	0.69
	Sumy	0.10	0.001	0.31	0.05	0.54	0.01
	Chernihiv	0.08	0.004	0.44	0.002	0.57	0.001
East	Donets'k	0.02	0.47	0.13	0.52	0.20	0.40
	Luhans'k	0.04	0.07	−0.03	0.91	0.08	0.77
	Kharkiv	0.11	0.004	0.35	0.02	0.59	0.01
Centre	Vinnytsia	0.002	0.87	−0.14	0.44	−0.17	0.40
	Dnipro	0.02	0.22	0.10	0.56	0.12	0.51
	Kropyvnyts'kyi	0.06	0.02	0.20	0.35	0.30	0.20
	Poltava	0.03	0.27	0.22	0.15	0.30	0.17
	Cherkasy	0.06	0.05	0.04	0.82	0.17	0.41
Ukraine		0.76	0.002	−0.10	0.06	−0.09	0.85

Table 5. Linear trend coefficient (α) and its significance (p) of recurrence (amount of events) and the snow deposit diameter (average and maximal) of the extreme soggy snow adhering in observation point in Ukraine and its regions in 1971–2010.

Soggy snow adhering in Ukraine is a very local phenomenon and occurs almost every year. The highest recurrence in observation point is characteristic for the West and South (2–3 times per year), especially on the south-west slopes of the Carpathians, in the Precarpathians and in the Crimea Mountains, where soggy snow adhering has a maximal average diameter of

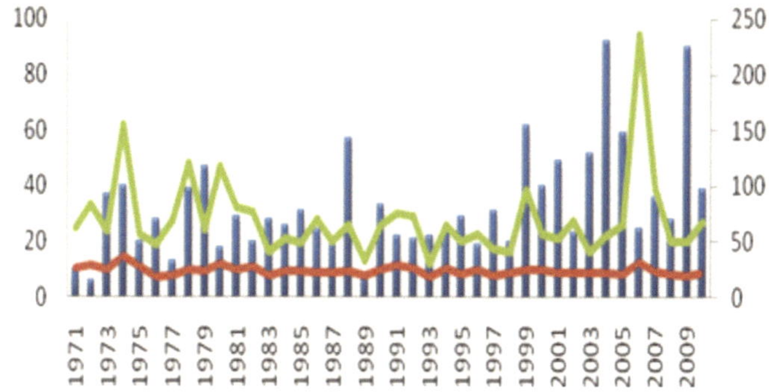

Figure 5. Changes of recurrence and snow deposits (diameter) of the extreme soggy snow-adhering events in Ukraine. Blue columns—amount of events. Red line—average diameter, mm. Green line—maximal diameter, mm.

25–30 mm and the maximal one can reach 200 mm and more. Eastern Polissia and north-western regions have the same recurrence but less diameter of 21–24 (average) and 120 mm (maximal). Recurrence in the central and south-eastern regions is once per year, the average diameter varies between 20 and 29 mm, and the maximal between 50 and 70 mm.

Because of low recurrence in one point of the phenomenon for the assessment of the long-term changes, both kinds of extreme soggy snow adhering (dangerous and distress) were analyzed. Analysis carried out has testified that in the period 1971–2010, the recurrence of soggy snow adhering and its diameter with 99% probability increased (**Table 5**, **Figure 5**) on 7–8 events in observation point for 10 years in average. The most significant increase of recurrence and diameter is observed in the North and East of Ukraine (Chernihiv, Sumy, Kharkiv regions). But increasing is characteristic not for the whole territory of Ukraine. A decrease is observed in the West, especially in Ivano-Frankivs'k region.

2.5. Compound (bedded) ice deposits

Compound ice deposits are ice layers formed on the object surface due to the sequential overlaying of different types of the ice which is caused by changeable weather conditions. The deposits consisted of glazed frost and dense snow-like friable atmospheric ice that are most dangerous. In Ukraine, such an event is dangerous when the compound ice deposits diameter is 11–34 mm and distress one when the diameter is >35 mm [8]. Such phenomena cause accident situations on the power engineering and communication lines, damage and losses in the transport, forestry, domestic economies, and so on. Compound ice deposits in Ukraine occur in the period from November to April (30% in December and 23% in January) in 6–9 o'clock in the morning and 18–20 o'clock in the evening. The diameter of the compound ice deposits is almost two times larger than the glazed frost. The average diameter fluctuates from 17 to 50 mm, the maximal one from 20 to 200 mm and more; 83% of the events have a diameter of <100 mm. In the West and South, this phenomenon is much more intensive. The share of >200 mm deposits is 2–7%. The total duration varies from a few hours to a few days and amounts to 40 h on average. The most long-continued compound ice deposits are in the western and southern regions. The maximal duration has been observed 27.11–5.12 2000 in

	Administrative region	Recurrence		Average diameter		Maximal diameter	
		α	p	α	p	α	p
West	Volynian	–0.067	0.067	–0.365	0.056	–0.355	0.085
	Transcarpathian	–0.037	0.016	–0.290	0.119	–0.315	0.107
	Ivano-Frankivs'k	–0.196	0.00002	–0.169	0.202	–0.333	0.026
	Lviv	–0.423	0.0006	–0.227	0.103	–0.393	0.023
	Rivne	–0.027	0.136	–0.235	0.186	–0.277	0.145
	Ternopil	–0.051	0.089	–0.142	0.452	–0.171	0.394
	Khmelnyts'kyi	–0.095	0.025	–0.526	0.002	–0.625	0.001
	Chernivtsi	–0.009	0.539	–0.300	0.108	–0.309	0.104
South	Zaporizhzhia	–0.124	0.004	–0.448	0.004	–0.494	0.004
	Crimea	–0.265	0.069	–0.034	0.066	–0.111	0.01
	Mykolaiv	–0.004	0.834	–0.048	0.792	–0.062	0.740
	Odesa	–0.014	0.726	0.019	0.922	0.019	0.923
	Kherson	0.237	0.002	0.259	0.010	0.399	0.002
North	Zhytomyr	–0.062	0.047	–0.497	0.006	–0.541	0.006
	Kyiv	–0.063	0.028	–0.217	0.253	–0.253	0.207
	Sumy	–0.026	0.024	–0.186	0.261	–0.200	0.237
	Chernihiv	–0.061	0.001	–0.639	0.0004	–0.684	0.0003
East	Donets'k	–0.093	0.054	–0.157	0.363	–0.217	0.269
	Luhans'k	–0.279	0.0002	–0.238	0.136	–0.359	0.043
	Kharkiv	–0.090	0.025	–0.271	0.127	–0.319	0.086
Centre	Vinnytsia	–0.094	0.042	–0.283	0.137	–0.344	0.089
	Dnipro	–0.080	0.032	–0.565	0.0003	–0.620	0.0004
	Kropyvnyts'kyi	–0.057	0.100	–0.223	0.238	–0.251	0.197
	Poltava	–0.043	0.008	–0.332	0.065	–0.355	0.054
	Cherkasy	–0.080	0.025	–0.569	0.002	–0.593	0.002
Ukraine		–2.082	0.001	–0.013	0.231	–0.106	0.028

Table 6. Linear trend coefficient (α) and its significance (p) of recurrence (amount of events) and the diameter (average and maximal) of the extreme compound ice deposits in observation point in Ukraine and its regions in 1971–2010.

Khmelnyts'kyi region (199 h) and 29.11–6.12 2000 in Chernivtsi region (Nova Ushytsia, 155 h). The highest recurrence (82%) is characteristic for the events continued less than 3 days, 56%—less than 24 h. The phenomena continued more than a week are very rare.

The highest amount of the events in one observation point is in the Transcarpathian, Donets'k, and Luhans'k regions——4–5 per 10 years (maximum on the south-west slopes of the Carpathians

http://dx.doi.org/10.5772/intechopen.77306

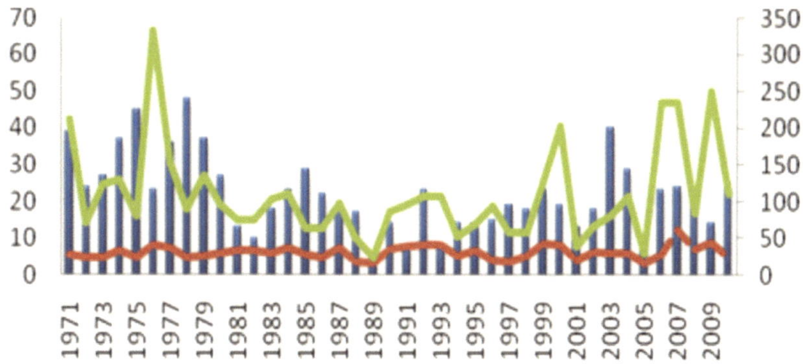

Figure 6. Changes of recurrence and the deposits (diameter) of the extreme compound ice deposits in Ukraine. Blue columns—amount of events. Red line—average diameter, mm. Green line—maximal diameter, mm.

and Donets'k Range). Significant recurrence (1–2 per 10 years) is observed in the Crimean Mountains, Ivano-Frankivs'k, Ternopil, and Kropyvnyts'kyi regions. In the northern regions, the compound ice deposits have not been observed in the analyzed period.

Compound ice deposits of the distress diameter are observed almost every year in the Crimea (Karabi-Yayla) and the Carpathians (Play) and occupy usually (99%) one district. There, the events are characterized by maximal diameter and duration. The maximal amount of days with the event (17) has been observed in 1970 in the South of Ukraine.

Because of low recurrence, the compound ice deposits of >11 mm were analyzed to assess their changes. The recurrence and diameter of compound ice deposits of distress diameter has been noticeably decreased with 99% probability (**Table 6** and **Figure 6**). Such trends are observed almost everywhere in Ukraine except the Kherson region. The most significant changes are observed in the northern and central regions, especially in the Chernihiv and Dnipro regions (**Table 6** and **Figure 6**).

2.6. Hail

In Ukraine, the hail belongs to a dangerous weather event, when its diameter is 6–20 mm and distress event, when its diameter exceeds 20 mm [8]. Most often (90%), the hail has a diameter of 20–50 mm. The hail of >50 mm diameter is observed very rare (6%) and in the southern regions usually. About 50% of events have a duration of 6–20 min. Hail continued more than an hour is observed very rarely. For example, on 22 July, 1987, in Sarat district of Odesa region, the hail has continued for 2 h 40 min and was a walnut size. Long-continued hail is observed in summer period usually and its duration in the mountain regions is higher than on the plains. The hail has a good marked daily variation with the maximum at 15–18 o'clock. In Ukraine, the extreme hail is observed in warm period usually. The recurrence of the extreme hail has also a good marked seasonal regime: maximum (70% occurrence) is observed in June and July. In spring and autumn, the extreme hail is rare. Mountain areas are characterized by a higher recurrence than plain.

From two to four events of distress, hail can be observed on 100 km^2 area for a 40-year period in Ukraine. Such hail can damage about 1–2% of agricultural areas. Within the whole

	Administrative region	Recurrence		Average diameter		Maximal diameter	
		α	p	α	p	α	p
West	Volynian	0.000	0.894	−0.054	0.750	−0.054	0.750
	Transcarpathian	−0.006	0.148	−0.207	0.124	−0.207	0.124
	Ivano-Frankivs'k	−0.010	0.060	−0.240	0.103	−0.265	0.085
	Lviv	−0.002	0.617	−0.022	0.878	−0.114	0.534
	Rivne	−0.001	0.285	−0.106	0.285	−0.106	0.285
	Ternopil	0.000	0.967	−0.020	0.907	−0.020	0.907
	Khmelnyts'kyi	−0.002	0.481	−0.083	0.463	−0.083	0.463
	Chernivtsi	−0.004	0.183	−0.058	0.183	−0.058	0.183
South	Zaporizhzhia	−0.002	0.510	0.052	0.764	0.036	0.836
	Crimea	−0.001	0.904	0.301	0.132	0.268	0.224
	Mykolaiv	0.007	0.020	0.296	0.056	0.329	0.044
	Odesa	−0.002	0.340	−0.430	0.090	−0.430	0.090
	Kherson	0.002	0.261	0.158	0.518	0.158	0.518
North	Zhytomyr	0.001	0.560	0.047	0.680	0.055	0.657
	Kyiv	0.002	0.464	0.221	0.341	0.222	0.341
	Sumy	0.003	0.161	0.206	0.092	0.206	0.092
	Chernihiv	−0.001	0.606	−0.054	0.617	−0.057	0.607
East	Donets'k	−0.002	0.132	−0.169	0.084	−0.169	0.084
	Luhans'k	0.001	0.451	0.107	0.388	0.107	0.388
	Kharkiv	0.001	0.662	−0.088	0.551	−0.072	0.639
Centre	Vinnytsia	−0.003	0.219	−0.144	0.279	−0.147	0.273
	Dnipro	0.001	0.764	0.042	0.812	0.051	0.779
	Kropyvnyts'kyi	0.003	0.402	0.101	0.385	0.135	0.363
	Poltava	−0.004	0.107	−0.207	0.126	−0.246	0.174
	Cherkasy	−0.002	0.568	0.011	0.940	0.011	0.940
Ukraine		0.000	0.545	0.092	0.445	0.065	0.830

Table 7. Linear trend coefficient (α) and its significance (p) of recurrence (amount of events per 100 km^2 per year) and the diameter (average and maximal) of the dangerous and distress hail in Ukraine and its regions in 1971–2010.

territory of Ukraine, these processes occur every year. The highest recurrence of distress hail is observed in the western regions (Transcarpathian, Ivano-Frankivs'k, and Ternopil) and the Crimea, where 6–7 distress hail events are observed on the 100 km^2 area for 40 years. The average diameter of the distress hail here varies from 23 to 29 mm, the maximum reaches 400 mm. Such hail can damage up to 20% of agricultural areas [11]. Southern slopes of the Volynian-Podolian, Dnipro, and Donets'k Highlands, which are opened for southern and

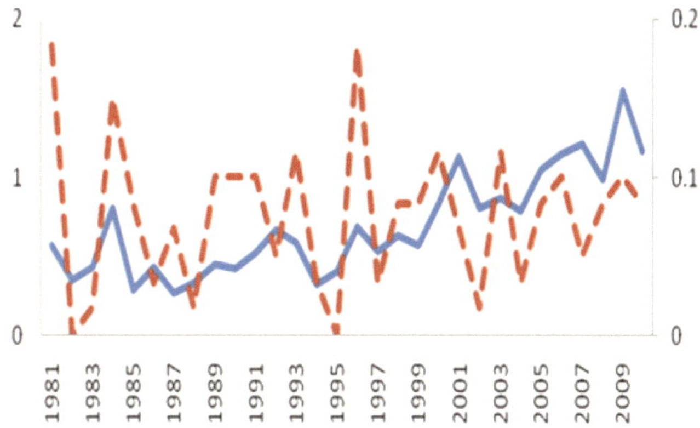

Figure 7. Changes of recurrence (event amount per 100 km^2 per year) of the dangerous and distress hail in Ukraine. Blue line—dangerous hail events. Red line—distress hail event.

south-western wet air masses, also are characterized by a significant distress hail recurrence (3–4 events per 100 km^2 per 40 years). But the hail intensity there is highest than in the West: the average diameter is 29–39 mm, and the maximum reaches 97 mm. The most dangerous in this aspect is the Bashtan district of Mykolaiv region, where not only the hail is very intensive but also other convective phenomena: squalls, storms, and spouts.

During the last decades, the tendency of the extreme hail recurrence increasing is observed in Ukraine. At that, the amount of hail with >6 mm diameter increases significantly, but with >20 mm it is not changed (**Table 7**, **Figure 7**). These changes were noticeably intensified since the end of the twentieth century because of convection processes intensity increase [11].

Some intensifying of the hail processes is observed in the Kropyvnyts'kyi, Dnipro, Kherson, and Mykolaiv regions where both recurrence and diameters of the hail have been increased. The average and maximal diameter has been increased also in the Crimea and Zaporizhzhia and Sumy region (**Table 7**).

2.7. Squalls

Among the dangerous weather phenomena of warm period, the share of squalls is not significant but the largest damages are caused by extreme squalls. In Ukraine, the squall with a wind velocity of 15–25 m/s is dangerous, with >25 m/s is distress. The average wind velocity of the extreme squalls changes from 18 to 22 m/s. Distress squalls have a wind velocity of 25–30 in 82.8% of events. A wind velocity of >40 m/s is very rare (1%). The most intensive squalls are observed in the south-western regions, where the average wind velocity amounts to 20–22 m/s, maximal—32–40 m/s. The share of squalls with 35 m/s velocity in these regions amounts to 20%, in the northern and southern regions—about 7%, in the central—4%.

The extreme squalls in Ukraine have a good marked seasonal recurrence. Most often (60%), they rise in June and July, 14% of events are observed in spring, about 7% in autumn, and about 3% in winter. Squalls have beaded daily variation with the maximum in the afternoon and evening hours (60% of events).

	Administrative region	Recurrence		Average velocity		Maximal velocity	
		α	p	α	p	α	p
West	Volynian	0.016	0.354	0.039	0.737	0.090	0.483
	Transcarpathian	0.032	0.002	0.286	0.013	0.317	0.011
	Ivano-Frankivs'k	0.006	0.462	0.143	0.297	0.153	0.291
	Lviv	0.061	0.003	−0.103	0.065	−0.079	0.386
	Rivne	0.021	0.017	0.157	0.224	0.194	0.149
	Ternopil	0.024	0.049	0.137	0.333	0.155	0.305
	Khmelnyts'kyi	0.057	0.009	0.221	0.093	0.284	0.050
	Chernivtsi	0.016	0.008	0.224	0.043	0.224	0.043
South	Zaporizhzhia	0.019	0.109	0.042	0.691	0.093	0.440
	Crimea	0.112	0.000	0.063	0.446	0.231	0.025
	Mykolaiv	0.030	0.003	0.301	0.012	0.375	0.005
	Odesa	0.046	0.013	−0.098	0.275	0.005	0.966
	Kherson	0.002	0.842	0.112	0.213	0.161	0.155
North	Zhytomyr	0.061	0.000	0.232	0.024	0.360	0.004
	Kyiv	0.058	0.000	0.219	0.029	0.336	0.009
	Sumy	0.047	0.000	0.316	0.014	0.370	0.009
	Chernihiv	0.005	0.560	−0.141	0.224	−0.213	0.117
East	Donets'k	0.010	0.201	0.083	0.492	0.121	0.380
	Luhans'k	0.053	0.000	0.352	0.003	0.438	0.001
	Kharkiv	0.058	0.000	0.233	0.039	0.383	0.003
Centre	Vinnytsia	−0.003	0.287	−0.152	0.293	−0.160	0.297
	Dnipro	0.039	0.000	0.276	0.045	0.414	0.008
	Kropyvnyts'kyi	0.014	0.120	−0.065	0.594	−0.065	0.634
	Poltava	−0.004	0.239	−0.178	0.172	−0.189	0.169
	Cherkasy	0.057	0.004	0.163	0.151	0.213	0.111
Ukraine		0.035	0.000	−0.041	0.005	0.032	0.563

Table 8. Linear trend coefficient (α) and its significance (p) of recurrence (amount of events per 100 km^2 per year) and the wind velocity (average and maximal) of the dangerous and distress squalls in Ukraine and its regions in 1971–2010.

Most of squalls (75% of events) continue less than 20 min, about 90% of them continue <30 min. Rarely, there are squalls of a 1-h duration. In the Odesa region 22.07.1987, the squall has continued more than 2 h with the wind velocity of 25 m/s and was accompanied by the hail of a walnut size. The largest amount of dangerous and distress squalls on the area of 100 km^2 per 40 years (1971–2010) has been observed in West and South, especially Lviv, Volynian, Odesa,

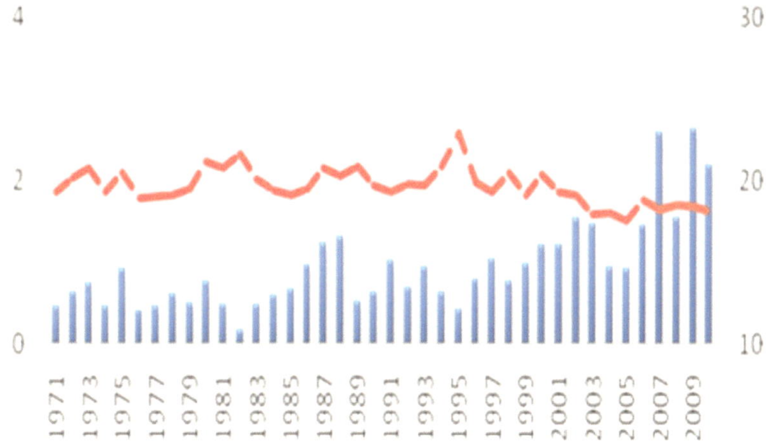

Figure 8. Changes of recurrence (event amount per 100 km^2 per year) and intensity of the dangerous and distress squalls in Ukraine. Blue columns—amount of events. Red line—average wind velocity, m/s.

Kherson, and Cherkasy regions (21–32 events) with an average wind velocity of 19–20 m/s, maximal—40 m/s. Significant recurrence and intensity are representative for the Polissia Lowland (except Rivne region), Khmelnyts'kyi, Zaporizhzhia, and Crimea (14–16 events). Extreme squalls are very rare in the eastern regions, where recurrence amounts to 3–7, an average wind velocity is 16–18 m/s, maximal—26–34 m/s.

The amount of the extreme squalls in Ukraine during the last decades has been increased. In the western and eastern regions, an increasing recurrence is accompanied by a significant increase in the squall intensity (average and maximal wind velocity). The same tendencies are representative for the Transcarpathian, Chernivtsi, Khmelnyts'kyi, Dnipro, Mykolaiv, and Crimea regions (**Table 8**, **Figure 8**).

2.8. Snowstorms

The snowstorms with a wind velocity of >15 m/s and a duration of >12 h are classified as distress in Ukraine [8]. For struggle with snow drifting on the roads, every year, millions of EURO are spent in Ukraine. Most of the snowstorms (70%) have a wind velocity of 15–30 m/s. This intensity events most often (95%) in the northern and central regions are observed. The more intensive snowstorms are representative in the western and southern regions, where the average velocity fluctuates from 25 to 29 m/s, the maximal one can reach 40 m/s and more. For instance, in the Ivano-Frankivs'k region, the share of snowstorms of 40 m/s amounts almost 25%. The least intensity snowstorms are representative for the North and North-East, where the average velocity is 17–21 m/s, maximal—24 m/s. Most of events (about 95%) last up to 2 days. The average duration is 15–30 h. The highest duration is observed in the Carpathian and Crimean Mountains, Donets'k Range. A maximal duration can reach 100 h. The absolute maximum has been fixed 10–20.02 1969 in the Luhans'k region (Daryivka)—242 h.

Snowstorm recurrence has a good marked yearly variation. The most intensive and long-continued snow-storms are observed from October to April with a maximum in February

(40.4%) and January (26.7%). About 12% of the events are observed in spring and 6% in autumn. Strong snowstorms can be observed any time of a day. Usually, they occupy a few administrative districts, but sometimes, they can occupy the area of a few regions. The snowstorms in Ukraine are observed almost every year. The highest recurrence is representative in the West, especially in the Ivano-Frankivs'k and Transcarpathian regions (2–9 events per 10 years), where the snowstorms have the highest wind velocity and duration. A significant recurrence is observed also in the southern and south-eastern regions. In the northern and central regions, the recurrence amounts 1–3 events per 10 years.

The snowstorm recurrence intensity decreases in Ukraine. The most significant changes are observed in the southern and northern regions, where recurrence and wind velocity have been decreased with 95% probability. At the same time, in the Ivano-Frankivs'k and Transcarpathian regions, especially in the mountains, the recurrence and wind velocity noticeably increase.

3. Conclusions

During 1971–2010, in Ukraine, the significant changes of the extreme weather phenomena recurrence and intensity have been observed which are caused by the meteorological preconditions of their forming, atmosphere circulation, and thermal regime changes.

In many regions of Ukraine, there is a trend of the rains and soggy snow recurrence increasing which has caused an increase of extreme soggy snow-adhering events amount. At the same time, the recurrence of the dangerous and distress glazed frost and compound ice deposits decreased. But since the start of the twenty-first century, an increase of these phenomena is observed.

During the last 40 years, also recurrence and intensity of the heavy snowfalls has been significantly increased. Also, the spatial occurrence of this phenomenon changed. In 1971–1990, the heaviest snowfalls occurred in the western and north-western regions, but during the last decades, the recurrence and intensity in these regions has decreased, simultaneously in the southern and south-eastern regions, they have been noticeably increased.

The air temperature, especially the maximal one in the warm period and air humidity increase, has caused convection intensification in Ukraine and unstability of the atmosphere. Due to such changes, the amount and intensity of the dangerous and distress rains, squalls and hail have increased.

Changes of macro- and mesoscale circulation in the second half of the twentieth century in the northern hemisphere have caused a decrease in strong baric gradient events which conditioned strong and long-continued winds. As a result, the recurrence of not only strong winds but also strong snowstorms and sand storms has been decreased.

The ascertained tendencies have regional specialties caused by the changes of their forming preconditions in these regions.

The next research consists in the assessment of the possible regional changes of extreme weather events for different scenario of the climate change, society development, and green gas emissions as well as social-ecological and socioeconomical effects of these changes.

http://dx.doi.org/10.5772/intechopen.77306

Author details

Vira Balabukh[1], Olena Lavrynenko[1], Volodymyr Bilaniuk[2], Andriy Mykhnovych[3]* and Olha Pylypovych[3]

*Address all correspondence to: 2andira@ukr.net

1 Department of Synoptic Meteorology, Ukrainian Hydrometeorological Institute, National Academy of Sciences of Ukraine, Kyiv, Ukraine

2 Department of Physical Geography, Ivan Franko National University of Lviv, Lviv, Ukraine

3 Department of Applied Geography and Cartography, Ivan Franko National University of Lviv, Lviv, Ukraine

References

[1] Lipins'kyi V, Osadchyi V, Babichenko V. Activization of the distress meteorological phenomena on the territory of Ukraine—Global climate change displaying. Ukrainian Geographic Journal. 2007;**2**:11-20

[2] Klein Tank AMG, Zwiers FW, Zhang X. Guidelines on analysis of extremes in a changing climate in support of informed decisions for adaptation. Climate Data and Monitoring. WCDMP-No 72. WMO-TD No 1500. 2009. 52 p

[3] Core Writing Team, Pachauri RK, Meyer LA, editors. IPCC: Climate Change 2014: Synthesis Report. Contribution of Working Groups I, II and III to the Fifth Assessment Report of the Intergovernmental Panel on Climate Change. Switzerland, Geneva: IPCC; 2014. 151 p

[4] Aleksander LV, Zhang X, Peterson TC, Caesar J, Gleason B, Klein Tank AMG, Haylock M, Collins D, Trewin B, Rahimzadeh F, Tagipour A, Rupa Kumar K, Revadekar J, Griffiths G, Vincent L, Stephenson DB, Burn J, Aguilar E, Brunet M, Taylor M, New M, Zhai P, Rusticucci M, Vazquez-Aguirre JL. Global observed changes in daily climate extremes of temperature and precipitation. Journal of Geophysical Research. 2006;**111**:D05109. DOI: 10.1029/2005JD006290

[5] Shevchenko O, Lee H, Snizhko S, Mayer H. Long-term analysis of heat waves in Ukraine. International Journal of Climatology. 2014;**34**(5):1642-1650

[6] Karl TR, Meehl GA, Miller CD, Hassol SJ, Waple AM, Murray WL, editors. CCSP: Weather and Climate Extremes in a Changing Climate. A Report by the U.S. Climate Change Science Program and the Subcommittee on Global Change Research. Washington, DC: NOAA's National Climatic Data Center; 2008. 164 p

[7] Osadchyi V, Babichenko V. Dynamics of the distress meteorological phenomena in Ukraine. Ukrainian Geographic Journal. 2012;**4**:8-14

[8] Meteorology. Terms and Definitions. Kyiv: State Standard of Ukraine; 1997. 57 p

[9] Sosnovs'ka R, Kulbida M, Humonenko L. Instructions for the Service on Prediction and Warning of Dangerous and Distress Hydrometeorological Phenomena and Weather Events. Official issue. Kyiv: State Hydrometeorological Service; 2003. 32 p

[10] Klein Tank AMG, Können GP. Trends indices of daily temperature and precipitation extremes in Europe, 1946-99. Journal of Climate. 2003;**16**:3665-3680

[11] Lipins'kyi V, Osadchyi V, Babichenko V, editors. Distress Meteorological Phenomena on the Territory of Ukraine during Last 20 Years (1986-2005). Kyiv: Nika-Centre; 2006. 312 p

[12] Logvinov K et al. Dangerous Weather Phenomena in Ukraine. Gidrometeoizdat: St. Petersburg; 1972. 236 p

[13] Babichenko V, editor. Extreme Meteorological Phenomena in Ukraine and Moldova. St. Petersburg: Gidrometeoizdat; 1991. 224 p

[14] Kinash R, Burnayev O. Extreme Hydrometeorological Phenomena in Ukraine. Lviv: STLP; 2000. 192 p

Chapter 6

Heat Waves: Health Effects, Observed Trends and Climate Change

Martínez-Austria Polioptro F and Bandala Erick R.

Additional information is available at the end of the chapter

http://dx.doi.org/10.5772/intechopen.75559

Abstract

According to climate change scenarios, the average annual temperature will increase by around 4°C if current trends continue. Maximum temperatures, however, have already registered higher values in different regions of the world, increasing the number, duration and intensity of heat waves. With the increase of maximum temperatures and the increase of significance of heat wave events, reports of mortality episodes due to heat effects have been increasing. According to the information from the Centre for Research on Epidemiology of Disasters (CRED), 5 of the 20 deadliest disasters between 1996 and 2015 were heat wave events. This chapter analyzes heat wave events, the criteria for determining dangerous temperature thresholds, as well as trends already observed, and those expected due to climate change. Heat wave events are correlated with observed increases on mortality rates, and recommendations are suggested to prevent their effects on human health.

Keywords: maximum temperatures, heat waves, heat and health, heat waves and climate change, heat wave mortality

1. Introduction

The number and cost of natural disasters related to climate have increased significantly in recent decades. Goklany [1] elaborated a study of the annual average number of climate-related extreme events, recorded per decade, during 1900–2008. He found a highly remarkable growth since the middle of the twentieth century; while 23 weather extreme disasters were observed in 1950–1959, and 354 were recorded only for the first decade of twenty-first century, until 2008.

According to Hoeppe [2], the number of events that caused human or material losses triplicated from 300 in 1980 to 900 in 2014. In fact, in the 2018 Global Risks Report of the World Economic Forum, the risk of extreme weather events shows the most unfavorable combination of impact and probability of occurrence, which poses the greatest global risk to humanity nowadays [3].

Heat wave events have become a major concern, mainly due to their effects on human health. According to the World Meteorological Organization (WMO), the number of casualties due to extreme temperature grew 2300% between 1991 and 2000 and 2001–2010 [4] being, by far, the greatest growing rate among climate-related disasters. This pervasive trend of heat wave events keeps going. For instance, the Centre for Research on Epidemiology of Disasters (CRED) have reported that 4 of the 10 deadliest disasters in 2015 were heat wave events [5] in agreement with one of the key findings of the Intergovernmental Panel for Climate Change (IPCC) Fifth Assessment Report saying that "changes in many extreme weather and climate events have been observed since about 1950. Some of these changes have been linked to human influences, including a decrease in cold temperature extremes, an increase in warm temperature extremes …" [6].

In the last few decades, records of particularly dangerous heat wave events causing several deaths, even in developed countries with good public health services, have been reported. For example, the heat wave at Chicago in July, 1995 causing 514 heat-related deaths [7], the extreme heat of 2003 in Europe, affecting mainly France, with almost 15,000 deaths [8] and the one at the Russian Federation in 2010, which caused 55,736 deaths [9].

Climate change is expected to increase the average temperature as well as the probability of extreme climate, including heat wave events [10–12]. Moreover, the best climate data estimates coming from the general circulation models are for average temperatures, making uncertainty in extreme temperatures and heat wave events much greater [13]. Predictions of temperature increases are based on results from the general circulation models. In this sense, the IPCC has estimated the performance of these models by comparing their results with climate parameters observed during 1980–1999. For temperature, once the multimodel results (the average of 23 general circulation models) have been analyzed, the estimated error (the difference between observed and modeled data) is rarely higher than 2°C, even though the individual models may show errors of up to 3°C [14]. Nevertheless, the IPCC has pointed out that large-scale trends are simulated with higher accuracy than regional ones. Therefore, the analysis of the vulnerability to and effects of climate change, at local or regional levels, must be based on observed evidence. Hence, local trend analysis is crucial for the analysis of heat wave events. Another reason, probably more important, for local analysis of heat wave events and extreme temperatures is that its effects on human health depend to a large extent on the level of acclimatization of the local population, which includes not only physiological aspects, but others like practices of prevention, transportation patterns, age and/or type of constructed environment (i.e. [8, 15]).

Heat wave events have not been sufficiently considered by most governments, while more attention and resources paid to other phenomena with greater public visibility, such as floods or severe storms, for example. Usually, the population is also not fully aware of the risk posed by extreme heat events. Therefore, any effort that spread heat wave events relevance, helps to mobilize governments, social organizations and individuals towards better prevention and a greater aware is of great importance; before, during and after a heat wave event.

2. Heat wave events: definition and thresholds

A heat wave is a period, usually lasting several days, with temperatures significantly higher than average or maximum values observed in the past during the same dates. The importance of extreme temperatures, and hence of heat wave events, becomes more significant when it affects human health. However, because the health-related effects of temperature depend on local factors—such as population acclimatization, age or health precondition—and other meteorological variables—such as humidity and wind speed—dozens of indices to determine the risk threshold of a heat wave have been reported. For instance, Smith et al. [16] listed 16 indices used in the United States.

A group of indices suggests the use of apparent temperature, which is calculated as a function of the observed temperature and relative humidity. The best known of them has been recommended by the US National Oceanographic Administration (NOAA), where the apparent heat index is calculated with the following equation:

$$\begin{aligned} H_i &= -42.379 + 2.049T + 10.14R - 0.224TR - 6.83 \times \\ &10^{-3}\, T^2 - 5.48 \times 10^{-2}\, R^2 + 1.22 \times 10^{-3}\, T^2 R + 8.52 \times \\ &10^{-4}\, T R^2 - 1.99 \times 10^{-6}\, T^2 R^2 \end{aligned} \tag{1}$$

where H_i is the apparent temperature, R is the relative humidity and T is the ambient temperature (°F). A H_i of 124°F or higher is considered very dangerous.

The main drawback of the heat index is related with the lack of access to relative humidity information. Recently, an extensive analysis to compare different heat indices with the registered health effects was carried out [17]. These authors found that simple indices based solely on temperature may be the most applicable for use in alert systems, but that all the corresponding temperature thresholds should be considered for regional analysis. Another group of threshold criteria establishes an absolute boundary to declare a heat wave, which is a specific value of the maximum temperature. So, for example, in the United Kingdom, the Met Office National Severe Weather Service (NSWSS) establishes a regional system with a threshold of maximum day and night temperatures, as shown in **Table 1**. In Mexico, a threshold for heat wave events in Mexico City has been proposed [18] as a maximum observed temperature greater than or equal to 30°C for three or more consecutive days, with an observed average temperature of 24°C. These authors suggested that, to determine the absolute temperature limits used in these criteria, it is necessary to identify the temperature where the mortality increases over average values and can be attributed to heat waves.

A second group of thresholds is defined by statistical analysis, usually selecting some temperature value percentile as the upper boundary. Some indices use the average daily

NSWWS Region	Temperature, °C	
	Day	Night
London	32	18
South East	31	16
South West	30	15
Eastern	30	15
West Midlands	30	15
East Midlands	30	15
North West	30	15
Yorkshire and Humber	29	15
North East	28	15

Table 1. Threshold maximum day and night temperatures defined by NSWSS [19].

temperature as reference temperature, while others—most of them—use maximum temperature values. As for extreme temperature events duration, some indices consider a heat wave when temperature exceeds the threshold limit during two or more days, while others estimate a heat wave occurs when the limit is exceeded in a single day.

The WMO Task Team on Definition of Extreme Weather and Climate Events encourages the report of the number of days in a month above percentile 90th [20]. In France, after the heat wave in 2003, a warning system was defined with an index based on the maximum and minimum temperatures, on average of 3 days, applicable to 96 cities of that country, based on studies in 14 pilot cities. Limits were determined by studying excess mortality caused by extreme heat, and the 98th percentile temperature value was used in the 2004 warning system. Subsequently, in 2005, the limits were updated for some cities, based on a more detailed analysis, and adopting the 99.5th temperature value percentile as the threshold. The resulting limits rank from 18 to 36°C for Lille to 21 to 36°C for Toulouse [21–23].

The methodology used in the warning systems of the United Kingdom and France, based on local studies—by city or region—for the relationship between temperature and excess mortality is interesting. Nevertheless, mortality information may not be available or the costs of these studies may not be suitable to many communities, and protecting the population through warning systems remains necessary. In these cases, it is advisable to use a boundary based only on maximum temperatures, with a conservative threshold used in national or international experience, for example, the 90th percentile, while the detailed studies are conducted.

It is worth to note that differences between definitions of heat wave events used in different countries represent a limitation to elaborate comparative analysis on the phenomenon between different regions of the planet.

3. Heat waves, morbidity and mortality

Extreme heat and effects on the population's health (higher morbidity and mortality) have been consistently reported in the last few decades [24–26]. As climate change effects become more significant, understanding and taking actions that may help to prevent heat-related health effects in the exposed population is an imperative topic that worth effort [27] in order to identify major trends, and vulnerable subpopulations since the frequency of these type of events is expected to increase in the near future [28, 29]. Some regions are expected to experience more intense, frequent and long-lasting extreme heat events by the second half of the twenty-first century with the corresponding health consequences.

While the scientific community remains discussing the best definition of a heat wave, such extreme weather-related events have continuously produced notable impact on human mortality, regional economies and ecosystems [29]. When the consequences of climate change are considered, warmer climate is expected to generate more intense, longer lasting and frequent heat waves contributing not only to human mortality but causing economic impacts, inconvenience and discomfort. The threat posed by extreme heat in a society not properly prepared to respond to weather-related events is significant considering that heat waves are 3 of the 10 worst natural disaster-related to the number of deaths recorded in 2013 [30].

Climate change sensitive areas are of particular concern mainly related with the effect of heat wave episodes on most vulnerable population subgroups such as children, the elderly, outdoor workers or inhabitants of urban areas [31]. Extreme heat events already account for the highest number of annual fatalities in the United States, and the annual number of heat waves has been found increasing for the average US city with increased length and intensity. Just to mention an example, Southwestern United States is a region of particular interest because it is predicted with increased average temperature and probability of extreme climate events, particularly in the second half of this century, based on results from general circulation models [32]. Among the different consequences, effects on population resulting of exposure to extreme heat range from dehydration, heat exhaustion, deadly heatstroke, kidney problems, lethargy and death; along with other side effects such as poor work attendance and/or performance among the exposed population.

In general, according to IPCC, regions in the middle and high latitudes will experience the highest temperature increases in the course of this century with a mean global temperature expected up to 4°C [6], based on the results of general circulation models. They have noticed, however, that large-scale features of climate are simulated with more accuracy that regional- and small-scale features, identifying the need of improved models capable of considering not only that changes in extreme events are not evenly distributed in space but characterized by particular patterns related to larger scale climate changes. Several of these places are projected also having the largest increase in population over the next decade and a significant increase in the frequency, last and intensity of extreme heat events [31] putting people and businesses at risk to the impacts of climate-related hazards. **Figure 1** shows the rising heat wave trends in the large US cities as an example of the overall tendency related with heat waves all around the world [32].

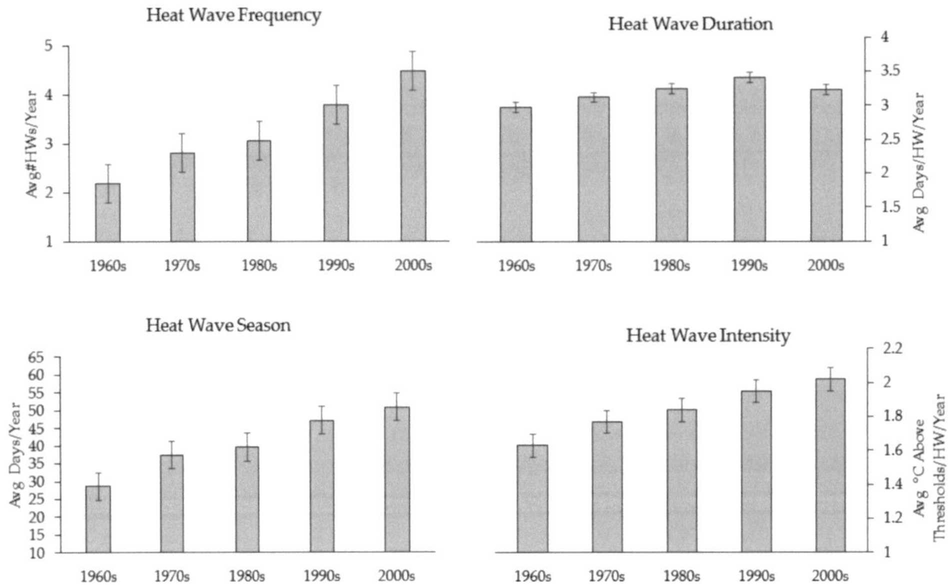

Figure 1. Decadal average trend for different heat wave characteristics across 50 US cities [32].

As shown, intensity on heat waves trends to increase as the frequency and heat wave season from the 1960s to 2000s, and only a slight decrease was identified for the duration of every heat wave event in the same period. The authors found the magnitude of the chance of the first three characteristics being rising by over 5% with the frequency of the heat wave increasing 20% per decade, generating undesirable effects on the population.

High body temperature is usually related with increase in heart and respiratory rate and, when reaching extreme values, also with brain, heart, lungs, kidney and liver damage. All these body responses to extreme heat can hardly be ignored, generating people to visit emergency rooms at hospitals and a significant amount of deaths every year. For example, only in the United States, more than 7400 people died from heat-related causes during 1999–2010 and over 65,000 visited emergency rooms each year [33, 34].

Table 2 shows the count of heat-health events accounted in California, USA between 2000 and 2009, including the excess hospitalization count and quantile. As shown, the count of excess hospitalization occurring statewide ranks as high as the frequency of heat wave event increases probably due to the exhaustion generated for the extreme heat events in the population organism [28].

Nevertheless, the trend in increasing the number of people suffering of heat-related effects is not privative of a specific location as suggested recently in a work where the temporal variation in heat-mortality associations in multi-country data set was assessed [35]. These authors found a significant prevalence of heat-related death in the studied period for the seven countries included in their study as shown in **Table 3**. Nevertheless, they found a decrease in the heat-related mortality risk in most of the countries analyzed and, in some cases, completely abated.

http://dx.doi.org/10.5772/intechopen.75559

Year	Event span	Peak date	Duration, days	Excess hospitalizations
2000	May 18–24	May 21	7	217
	Jun 12–16	June 14	5	299
	Sept 5–9	Sept 7	5	700
2001	May 2–11	May 8	10	959
	May 29–Jun 1	May 31	4	460
2002	Jun 7–13	Jul 9	7	848
2003	May 19–22	May 20	4	845
	May 27–29	May 28	3	454
	Jun 24–30	Jun 27	7	717
	Jul 8–22	Jul 14	15	1063
	Sept 10–15	Sept 13	6	629
	Sept 17–23	Sept 22	7	839
2004	Sept 1–17	Sept 7	17	1657
2006	Jul 7–9	Jul 8	3	367
	Jul 13–26	Jul 23	14	1254
2007	May 6–9	May 7	4	327
2008	May 13–18	May 16	6	903
2009	May 15–18	May 17	4	160
	Aug 7–17	Aug 10	11	228

Table 2. Heat-health events in California, US in 2000–2009 (taken from [28]).

Country	Total deaths	Period	Summer temperature, °C		
			Minimum	Median	Maximum
Australia	361,135	1988–1998	14.5	21.6	32.2
		1999–2009	14.8	22.1	32.7
Canada	679,693	1986–1998	3.9	17.4	28.0
		1999–2011	4.4	17.7	28.5
Japan	8,117,084	1985–1998	14.9	24.3	31.7
		199–2012	15.3	25.1	31.6
South Korea	530,618	1992–2001	14.9	23.7	31.8
		2002–2010	14.5	23.6	30.5
Spain	1,050,433	1990–2000	12.4	22.2	31.1
		2001–2010	13.3	22.6	31.5

Country	Total deaths	Period	Summer temperature, °C		
			Minimum	Median	Maximum
UK	2,285,519	1993–1999	8.5	15.3	24.3
		2000–2006	11.1	23.6	31.8
USA	6,994,609	1985–1995	11.1	23.6	31.8
		1996–2006	11.7	23.6	31.9

Table 3. Heat-related deaths in seven countries during the last two decades (taken from [35]).

Despite the controversial results, it is clear that exposure to extreme heat poses a significant threat to human health, mainly when pre-existing health conditions are present. The lack of information on the source of the risk or the physiologic response generated by extreme heat in the different individuals has been identified as a major gap in knowledge that worth to be filled [36] as well as the considerations for the proper diagnose of the heat-related effects on health.

4. Observed trends and climate change

The maximum temperature has shown a growing trend in recent decades. WMO has reported a steady increase in the number of countries reporting new maximum temperature records. As shown in **Figure 2**, while less than 10% of the countries reported new temperature records in 1961–1970, 44% (56 countries) reported new maximum temperature in 2001–2010 [4].

Most climate change publications have focused on average temperature values. However, a slight increase in the average temperature means, in a changing climate, a greater increase in maximum temperature values. Moreover, it can happen that average temperature does not change or even decrease while the maximum temperatures increase significantly. The explanation of this phenomenon is found in the possible changes in temperature distribution of probabilities, as indicated by the IPCC (see **Figure 3**). Changes in the average temperature can

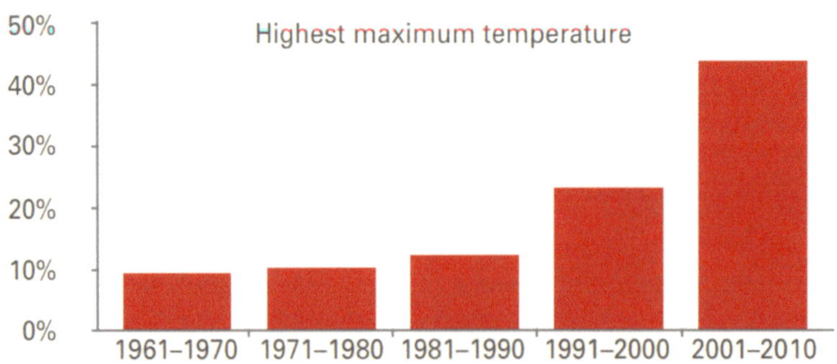

Figure 2. Absolute country records of the daily maximum temperature [4].

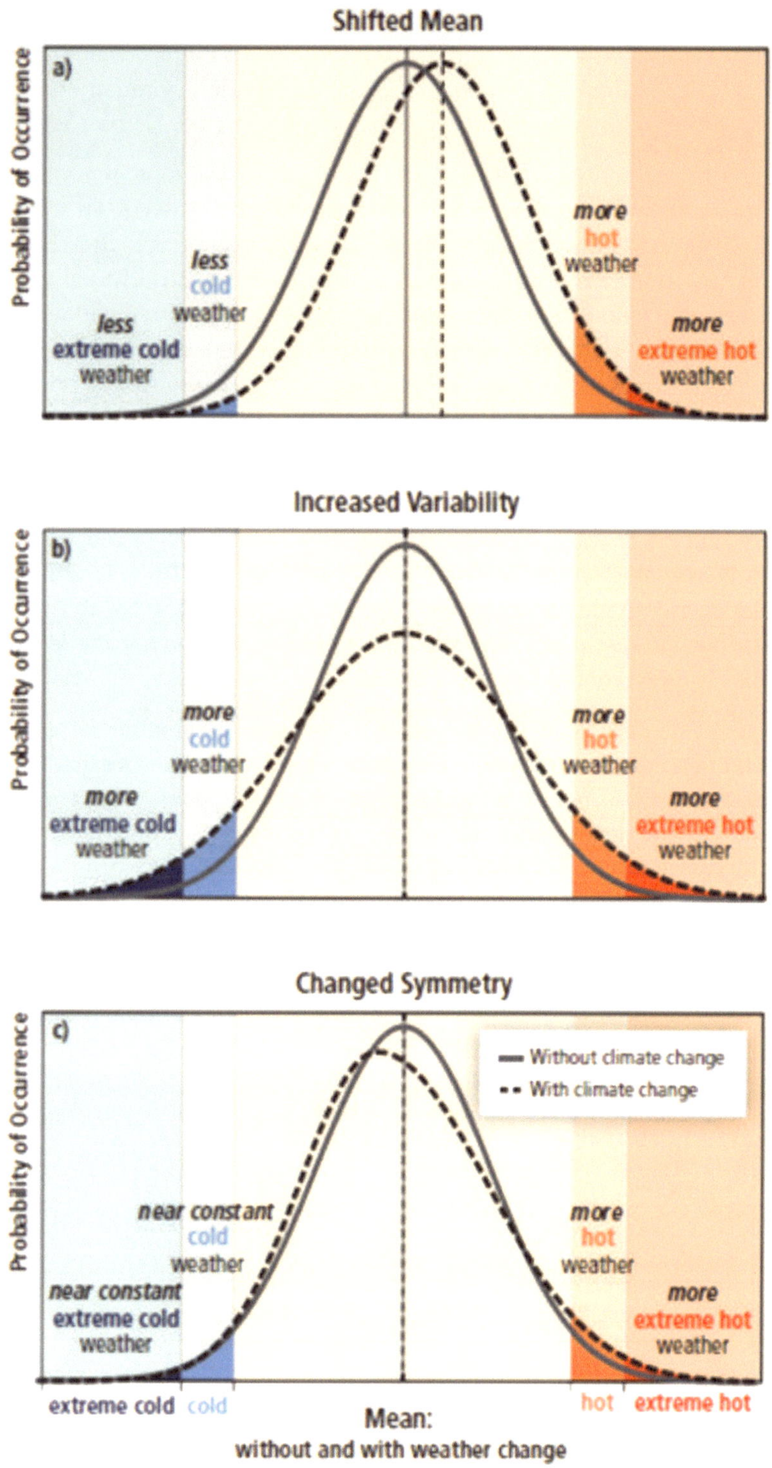

Figure 3. Effect of changes in temperature distribution extremes [37].

be produced, with a shift in the probability distribution toward areas of higher temperature (**Figure 3a**). An increase in variability may also occur, with more extreme temperatures—both, cold and heat (**Figure 3b**)—or changes in the distribution symmetry with decrease in average temperature and increase in temperature at both extremes (**Figure 3c**).

To illustrate this phenomenon, **Figure 4** shows the observed variation of minimum, medium and maximum temperatures in Mexicali (Northwestern Mexico) where an analysis on average temperature would indicate that the climatic change direction is to decrease, although in fact, it warms up and is increasingly higher for both diurnal and nocturnal values. Because minimum temperatures increase faster than maximum temperatures, average temperature value seems to decrease. In this case, changes in the symmetry of distribution probabilities are assumed. Therefore, the study of heat wave trends should be based mainly on maximum and minimum temperature analysis rather than average temperature values.

The IPCC special report on the risk of extreme events clearly states that "is very likely that there has been an overall decrease in the number of cold days and nights, and an overall increase in the number of warm days and nights" [37]. On the other hand, the magnitude of the change in the extreme temperatures depends on the region. There is sufficient evidence of significant increases in maximum temperatures, day and night, in North America, Europe and Australia. For the rest of the regions, it is necessary to consider the lack of data and detailed studies before a conclusion can be stated.

In the United States, a detailed study showed that in most of the continental territory, there is a positive trend in temperature for 1979–2011 with a level of significance greater than 95% in the Mann Kendall test [16]. These authors indicate that the regions of the United Sates with the greatest positive tendency are found in the Southeast and the Great Plains. These results are complemented by a study in the Southeastern United States and Northwest Mexico—the

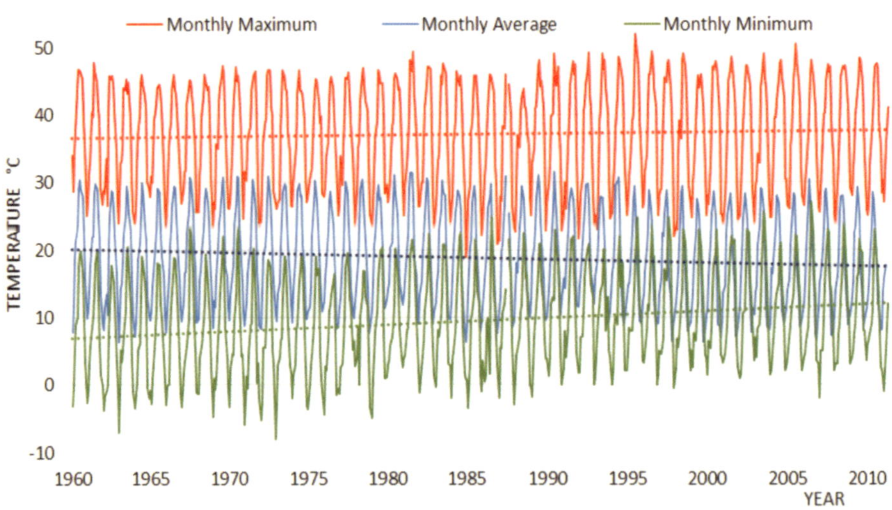

Figure 4. Maximum, average and minimum temperature trends observed in Mexicali, Mexico.

Sonoran and Mojave Desert—where a positive trend, using the Spearman's Rho test, during 1960–2010 in maximum temperatures of Tucson and Phoenix in the United States and Obregon City and Mexicali in Mexico was found [31]. As shown in **Figure 5**, there is a clear trend of increase in maximum temperatures throughout the region, clearly seen by comparing against average temperature over the analyzed time interval. During 2005–2010, the maximum temperature in Ciudad Obregon was 2.96°C higher than 1961–1965, 2.76°C higher in Phoenix, 2.52°C higher in Tucson, 2.3°C higher in Morelos Dam, 2.06°C higher in Mexicali and 1.34°C higher in Las Vegas. The number of days that exceeds the 90th maximum temperature percentile were found also growing fast, as shown in **Figure 6**.

Europe has been subject to exceptional heat wave events. The 2003 heat wave in Europe, widely documented, caused more than 33,000 deaths, and extended with record temperatures in the range of 35 to >40°C, from Portugal and England to Turkey [38]. France, the most affected country, reported more than 14,000 deaths [8]. During the 2010 heat wave in the Russian Federation, which caused more than 55,000 deaths, the temperature anomaly in Moscow was as high of +18°C in July [39]. In the United Kingdom, a definite trend in the increase of moderate heat wave events, both in number and duration, while the trends for extreme heat wave events are still unclear [40].

Australia is one of the countries that suffers most due to heat wave events. A report from the Australian Climate Council has found a notable increase in the magnitude, duration and frequency of heat waves. It has been found, for example, that in the last 50 years, the number

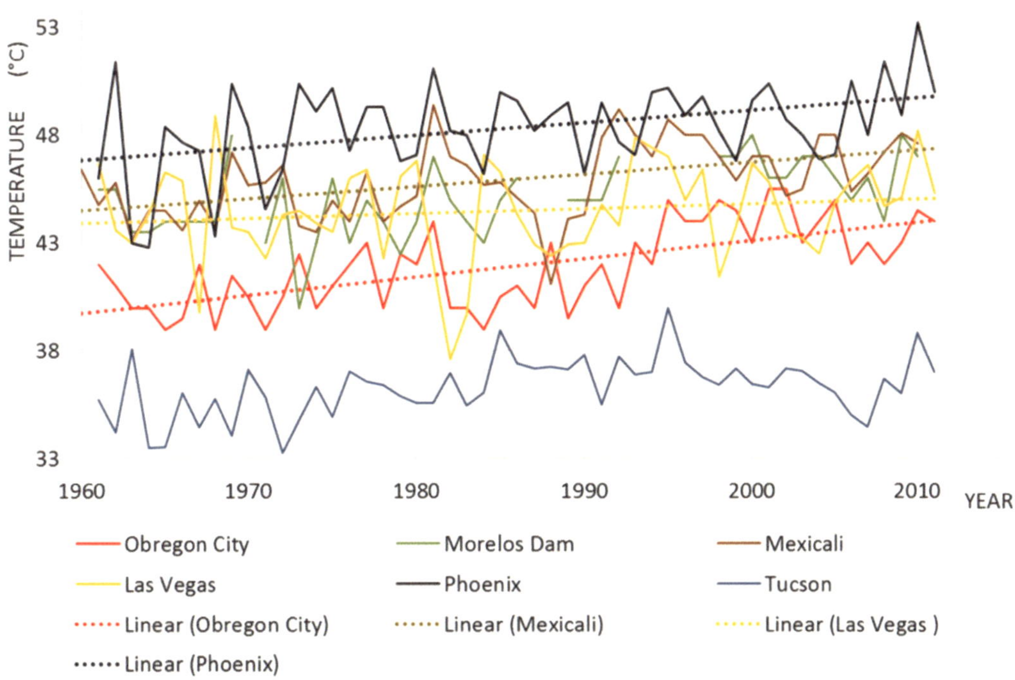

Figure 5. Maximum monthly temperatures and linear trends for the Sonoran Mojave Desert region for the month August [31].

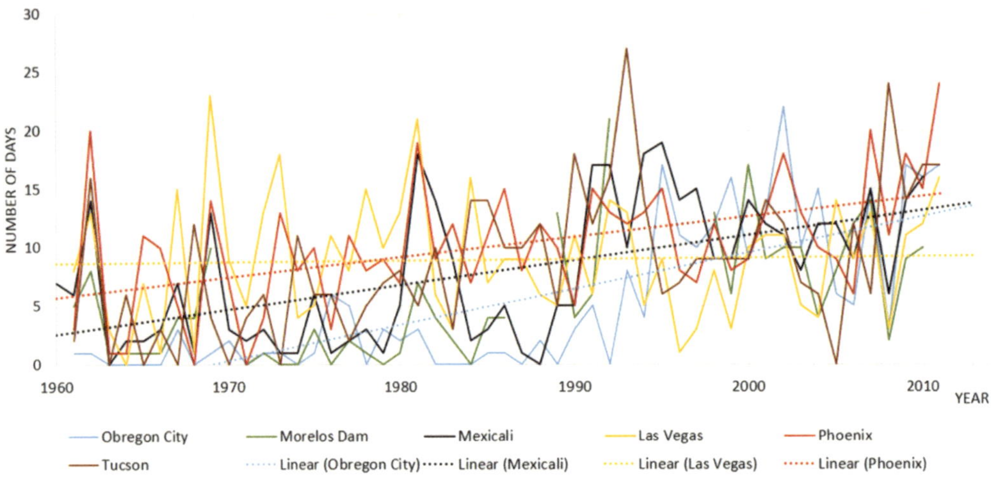

Figure 6. Number of days that exceeded the 90th percentile of average maximum temperatures in August, in the Sonoran Mojave Desert region [31].

of hot days in Australia has doubled and that "in the last hundred years, heat waves have caused more deaths than any other natural risk" [41]. Trend analyzes conducted so far show an increase in heat wave events' intensity, duration and frequency [42].

There are reports from many other regions of the world, which report the same growing trends in the number and duration of heat waves [43–45]. However, it is essential to emphasize the remaining lack of information and analysis in large areas, for example, at Central America, South America and Africa.

The trends observed in extreme heat and heat wave events, unfortunately, are in the direction predicted by climate change scenarios. Indeed, the Fifth Report of the IPCC indicates that during the twenty-first century, "it is very likely that heat waves will occur more often and last longer" [6]. The special report on extreme events of the IPCC is even more conclusive, and states that "it is virtually certain that increases in the frequency and magnitude of warm daily temperature extremes and decreases in cold extremes will occur in the twenty-first century at global scale" [37]. The magnitude of these events at the regional level, however, will depend on local variables such as elevation, synoptic effects, the occurrence of anticyclones, and so on, which is why more precise modeling at smaller scales is essential as well as their calibration according to the changes will develop over time.

For Europe, the results of a simulation model for the twenty-first century, with a mesh resolution of 50 km were recently published [46]. The report shows a highly significant increase in the number of days with heat wave events per year, the maximum average temperature and the frequency. From the trends in their report, the number of heat wave events per year will increase approximately, during 2000–2100, from 1 to 2.5 in Western Russia, from 10 to 27 in Eastern Europe and from 1 to almost 2.5 in Western Europe. The mean maximum temperature, in the same period, will grow from 19 to >24°C in Western Russia, from 20 to almost 26°C in Eastern Europe and from around 18 to >24°C in Western Europe [46].

	Northern Europe		Eastern Europe		Central Europe		Western Europe		Southern Europe	
	1981–2000	2071–2100	1981–2000	2071–2100	1981–2000	2071–2100	1981–2000	2071–2100	1981–2000	2071–2100
Deaths /10^7 inhabitants	0.98	27.24	4.81	177.6	61.51	2305.77	34.33	1022.67	105.77	7019.99
Exposed population /10^6 inhabitants	9618	269,167	9989	341,025	9932	322,309	9993	298,558	9828	634,474

Table 4. Prognosis of people affected and deaths in Europe by heat waves (with data from [47]).

Forzieri et al. [47] analyzed the impacts on the European population due to climate change, in a period of 30 years. With respect to heat waves, **Table 4** presents their results of the changes in exposed population and expected deaths during 1981–2000 and 2071–2100.

For the United States, Kunkel et al. [48] estimated, for heat wave events in summer, that the annual 3-day heat wave temperature increased by 3–8°C and the number of heat wave days from 30 to 60 days per year. The analysis for other regions and countries also coincides in forecasting a growth of heat waves during the twenty-first century. For example, the study by Meehl and Tebaldi [29] can be cited who consider North America and Europe; the report by Steffen et al. [41] for Australia; another by Dutta and Chrsiya [49], focused on the effects on health due to climate change in India; or the projections of extreme heat performed by Salinas-Prieto et al. [50] for Mexico, forecasting increases between 4.6 and 5.8°C in maximum temperatures for northern Mexico during 2075–2099, using the RCP8.5 Scenario. Even since there is a consensus on the increase in intensity, frequency and duration of heat wave events due to climate change, there is still a great uncertainty about its magnitude at regional scale, so the study of the observed trends should be continued, and carried out where the information or analysis capabilities have not allowed it until now.

5. Concluding remarks

It seems evident that extreme heat, resulting from climate change, poses a significant treat not only for people's health, but also in general for achieving the Millennium Sustainable Development Goals proposed by the UN by being within the top dangerous climate-related disasters with the greatest gap in knowledge.

The lack of consensus on the definition and threshold values for heat wave events among the scientific community is beyond the academic work as it may become a life or death issue for vulnerable population segments or regions. As a result, the urge for generating more reliable information and used for decision making becomes with a higher significance as the time goes by.

The relationship among extreme heat during heat wave events and the amount of deaths or the increase of people visiting emergency room in hospitals is undeniable, but the scientific work reported nowadays does not suffice to figure out what consequences may occur in other regions where climate variability is barely known or completely unknown. The need for more studies is clear, and additional resources to be invested in such sensitive knowledge areas seem mandatory in order to avoid potential catastrophic scenarios in the forthcoming years.

Author details

Martínez-Austria Polioptro F[1*] and Bandala Erick R.[2]

*Address all correspondence to: polioptro.martinez@udlap.mx

1 UNESCO Chair on Hydrometeorological Risks, University of the Americas Puebla, Mexico

2 Division on Hydrological Sciences, Desert Research Institute, Las Vegas, Nevada, U.S.A.

References

[1] Goklany IM. Deaths and death rates from extreme weather events: 1900-2008. Journal of American Physicians and Surgeons. 2009;**14**(4):102-109

[2] Hoeppe P. Trends in weather related disasters -consequences for insurers and society. Weather and Climate Extremes. 2016;**11**:70-79

[3] World Economic Forum. The global risks report 2018, 13th ed. Geneva: World Economic Forum; 2018

[4] World Meteorological Organization, The Global Climate 2001-2010. A Decade of Climate Extremes. Summary Report. Geneva, Switzerland, 2013

[5] Guha-Sapir D, Hoyois P, Below R. Annual Disaster Statistical Review 2015. The numbers and Trends. CRED, Brussels, 2016

[6] IPCC, Climate Change 2014. Synthesis Report. Intergovernmental Panel of Climate Change. Oxford; 2014

[7] Withman S, Good S, Donoghue ER, Benbow N, Shou W, Mou S. Mortality in Chicago attributed to the July 1995 heat wave. American Journal of Public Health. 1997;**87**(9):1515-1518

[8] Kovats RS, Ebi KL. Heatwaves and public health in Europe. European Journal of Public Health. 2006;**16**(6):592-599

[9] World Meteorological Organization. Atlas of mortality and economic losses from weather, climate and water extremes (1970-2012). Geneva. 2014

[10] IPCC, Summary for Policymakers. In Climate Change 2013: The Physical Basis. Stocker TF, editor. Cambridge: Cambridge University Press; 2013

[11] Meehl GA, Tebaldi C. More intense, more frequent, and longer lasting heat waves in the 21st century. Science Reports. 13 August 2004;**305**:994-997

[12] Hughes L, Perkins S. Heatwaves: Hotter, longer, more Often. Sydney, Australia: Climate Council of Australia; 2014. p. 62

[13] Ganguly AR, Steinhaeuser K, Erickson DJ, Branstetter M, Parish ES, Singh N, Drake JB, Lawrence B. Higher trends but larger uncertainty and geographic variability in 21st century temperature and heat waves. PNAS. 2009;**106**(37):15555-15559

[14] Randall DA, Word RA. Climate models and their evaluation. In: IPCC, Climate 2007: Impacts, Adaptation and Vulnerability. Cambridge: Cambridge University Press; 2007. pp. 589-662

[15] World Meteorological Organization and World Health Organization, Heatwaves and Health. Guidance on Warning-System Development, Vols. WMO-No. 1142, McGregor GR. ed, Geneva Switzerland, 2015

[16] Smith TT, Zaitchik BF, Dohlkr JM. Heat waves in the United States: Definitions, patterns and trends. Climatic Change. 2013;**118**(3-4):811-825

[17] Kent ST, McClour LA, Zaitchik BF, Smith TT, Gohlke JM. Heath waves and health outcomes in Alabama (USA); the importance of heat wave definition. Environmental Health Perspectives. 2104;**122**(2):151-158

[18] Jáuregui E. The heat spells of Mexico City. Investigaciones Geográficas. 2009;**70**:71-76

[19] Public Health England, Heatwave Plan for England. London: NHS England2015

[20] WMO Task Team on Definition of Extreme Weather and Climate Events, Meeting Report 26-26 February. WMO, 2014

[21] Wagner V, Le Tertre A, Laaidi K. French heat health warning system: Validation of temperature thresholds. Institute de veille Sanitaire; 2006

[22] Pascal M, Laaldi K, Ledrans M, Baffert E, Caserlo-Sconemann C, Le Tertre A, Manach J, Medina S, Rudant J, Empereur-Bissonnet P. France's heat health watch warning system. International Journal of Biometeorology. 2006;**50**(3):144-153

[23] Laadi K, Ung A, Wagner V, Buaudeau P, Pascal M. The French Heat and Health Watch Warning System: Principles, Fundamentals and Assessment. Institute de veille sanitaire; 2013

[24] Kingsley S, Eliot M, Gold J, Vanderslice R, Wellenius G. Current and projected heat-related morbidity and mortality in Rhode Island. Environmental Health Perspectives. 2016;**124**(4)

[25] Basu R, Samet J. Relation between elevated ambient temperature and mortality: A review of the epidemiologic evidence. Epidemiologic Review. Vol. 24. 2002

[26] Petkova E, Gasparrini A, Kineey P. Heat and mortality in New York City since the beginning of the 20th Century. Epidemiology. Vol. 25. 2014

[27] Gubernot D, Anderson G, Hunting K. Characterizing occupational heat-related mortality in the United States, 2000-2010: An analysis using the census of fatal occupational injuries database. American Journal of Industrial Medicine. 2015;**58**(8)

[28] Guirguis K, Gershunov G, Tardy A, Basu R. The impact of recent heat waves on human health in California. Journal of Applied Meteorology and Climatology. 2014;**53**(1)

[29] Koppe C, Kovats S, Jendrotzky G, Menne B. Heat waves: Risks and responses. World Health Organization. Europe. Health and Global Environmental Change series, No.2. 2004

[30] Martinez-Austria P, Bandala E, Patino C. Temperature and heat waves trends in North-west Mexico. Physics and Chemistry of the Earth. 2016;**91**:20-26

[31] Martinez-Austria P, Bandala E. Temperature and heat-related mortality trends in the Sonoran and Mojave Desert region. Atmosphere. 2017;**8**(3)

[32] Habeeb D, Vaego J, Stone B. Rising heat wave trends in large U.S. cities. Natural Hazards, Springer Netherlands. 2015;**76**

[33] Hess J, Saha S, Luber G. Summertime acute hear illness in U.S. emergency department from 2006 through 2010: Analysis of a nationally representative sample. Environmental Health Perspectives. 2014;**122**:1209-1215

[34] Xu J. Morbidity and Mortality Weekly Report: Quickstats Number of Heat-Related Deaths, by Sex-National Vital Statistical System, United States, 1999-2010. Atlanta, GA: Centers for Disease Control and Prevention; 2012

[35] Gasparrini A, Guo Y, Hashizume M, Kinney P, Petkova E, Lavinge E, Zanobetti A, Schwartz J, Tobias A, Leone M, Tong S, Honda Y, Kim H, Armstrong BG. Temporal variation on heat-mortality associations: A multicountry study. Environmental Health Perspectives. 2015;**123**(11):1200-1207

[36] Scott R. Between extremes: Health effect of heat and cold. Environmental Health Pers-pectives. 2015;**11**:A275-A280

[37] IPCC, Managing the Risks of Extreme Events and Disasters to Advance Climate Change Adaptation. Field C, Barros V, Stocker TF, Dahe Q. editors, New York: Cambridge University Press, 2012

[38] A. De Bono, P. Peduzzi, S. Kluser and G. Giuliani, Impacts of summer 2003 heat wave in Europe. United Nations Environment Programme, 2004

[39] Grumm RH. The central European and Russian heat event of July-august 2010. Bulletin American Meteorological Society. 2011;(October):1285-1296

[40] Sanderson MG, Economou T, Salmon KH, Jones SE. Historical Trends and Variability in Heat Waves in the United Kingdom. Atmosphere, no. September. 2017

[41] Steffen W, Hughes L, Perkins S. Heatwaves: Hotter, longer, more often. Melbourne: Climate Council of Australia; 2014

[42] Perkins-Kirkpatrick S, White C, Alexander L, Argüeso D, Boschat G, Cowan T, Evans J, Ekström M, Oliver E, Phatak A, Purich A. Natural hazards in Australia: Heatwaves. Climate Change, no. Special Issue on "The effect of historical and future climate changes on natural hazards in Australia". 2016

[43] Arora M, Goel N, Singh P. Evaluation of temperature trends over India. Hydrological Sciences Journal. 2005;**50**(1):81-93

[44] Chaidez V, Dreano D, Agusti S, Duarte C, Hoteit I. Decadal trends in Red Sea maximum surface temperature. Nature-scientific reports. 2017;**15**(August)

[45] Schubert S, Wang H, Koster RD, Suárez MJ, Groisman PY. Northern Eurasian heat waves and droughts. American Meteorological Society. 2014:3169-3207

[46] Lau N, Nath MJ. Model simulation and projection of European heat waves in present-day and future climates. Journal of Climate. 2014;**27**(May):3713-3730

[47] Forzieri G, Cescatti A, Batista e Silva F, Feyen L. Increasing risk over time of weather-related hazards to the European population: a data-driven prognostic study. Lancet Planet health. 2017;**1**(August):e200-e208

[48] Kunkel KE, Liang X-Z, Zhu J. Regional climate model projections and uncertainties of U.S. summer heat waves. American Meteorological Society. 2010;**23**:4447-44458

[49] Dutta P, Chorsiya V. Scenario of climate change and human health in India. International Journal of Innovative Research and Development. 2013;**2**(8):157-160

[50] Salinas Prieto JA, Colorado Ruiz G, Maya Magaña ME. Climate Change Scenarios for Mexico (In Spanish), in Atlas de vulnerabilidad hídrica de México ante el cambio climático. Jiutepec, Morelos: Instituto Mexicano de Tecnología del Agua; 2015. pp. 41-70

Chapter 7

Malawi's Experience with Weather Index Insurance as Agricultural Risk Mitigation Strategy Against Extreme Drought Events

Ephias M. Makaudze

Additional information is available at the end of the chapter

http://dx.doi.org/10.5772/intechopen.77106

Abstract

Malawi continues to face unprecedented challenges imposed by extreme weather events—drought in particular. Because the economy is heavily dependent on climate-sensitive agriculture, Malawi is highly vulnerable to extreme drought events. Faced with a growing number of extreme drought events, the government of Malawi is determined to implement weather index insurance as part and parcel of its nationwide disaster risk mitigation strategy. This study seeks to interrogate and highlight the obstacles that have hampered successful implementation and development of weather index insurance in Malawi beyond the pilot phase. The study draws on a few examples to demonstrate other countries' successful experience with weather index insurance as a drought risk mitigation tool. The study concludes by highlighting important lessons that could help the government of Malawi to re-think the next phase of development of weather index insurance program as an effective risk protection strategy against extreme drought events.

Keywords: weather index insurance, extreme drought, smallholder farmers, agricultural insurance, climate change, insurance literacy, disaster risk reduction

1. Background

Malawi, like many other countries in Africa, continues to face severe challenges imposed by extreme weather events particularly droughts. The upsurge of drought events in Malawi provides evidence to unprecedented challenges caused by climate change and climate variability. Because the economy is heavily dependent on climate-sensitive agriculture, Malawi's economy is highly vulnerable to extreme weather events especially drought. Pauw *et al.* [1]

estimated that on average Malawi loses 1.7% of GDP (equivalent to USD 41 million) every year due to an extreme drought event.

In response to severe challenges imposed by drought, the government of Malawi, in partnership with the World Bank and other development partners, initiated a weather index insurance (WII) scheme in 2005—the first of this kind in Southern Africa. The scheme was 'celebrated' as a potential breakthrough in the quest to offer sustainable market-based agricultural insurance to millions of smallholder farmers—a majority who practices dryland rain-fed agriculture.

Like other developing countries in African, Malawi's interest in WII is justified on a number of factors: first, currently agricultural insurance markets are missing, fragmented or poorly developed. Because agriculture is the dominant source of livelihood to millions of smallholder farmers, WII has the potential to provide effective risk-protection mechanism against the often highly disruptive drought events. Second, because the economy is agro-based, agriculture is the most feasible strategy to promote economic growth in Malawi with knock-on effects on food security and poverty improvement. World Bank [2] assets that growth in agriculture is a more effective strategy at reducing poverty compared to

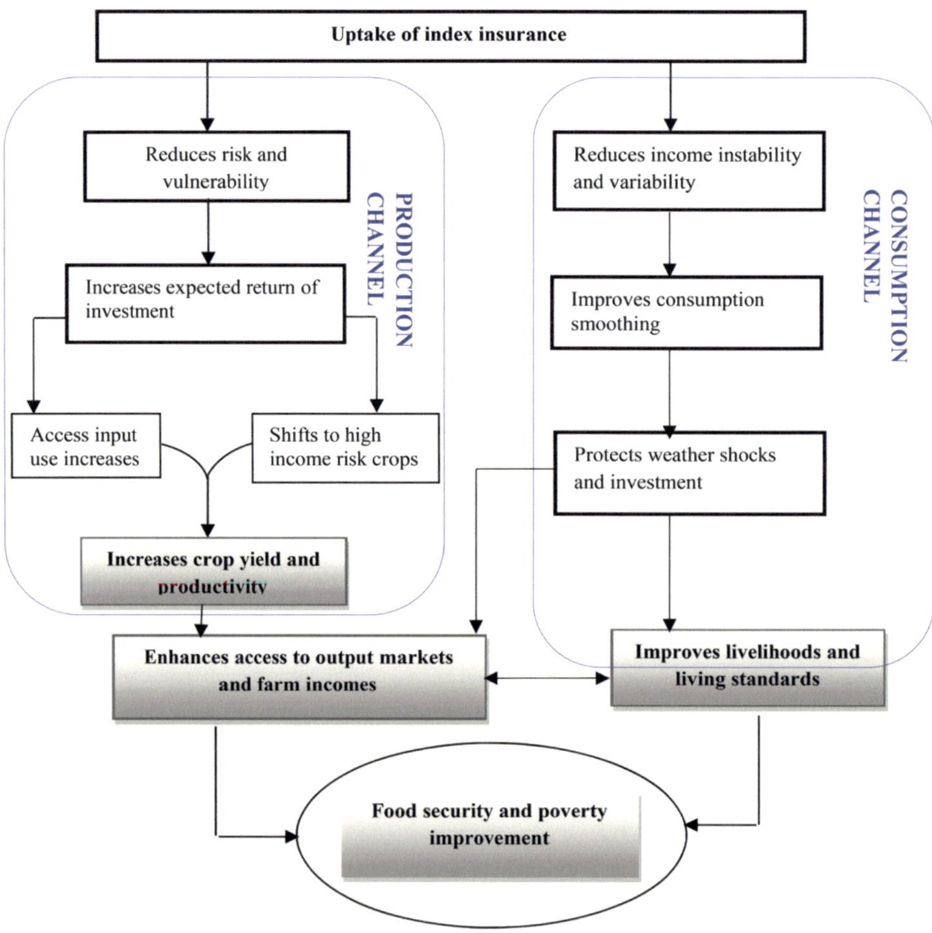

Figure 1. Rationale for uptake of WII by smallholder farmers. Source: Adapted from Cole et al. [5].

http://dx.doi.org/10.5772/intechopen.77106

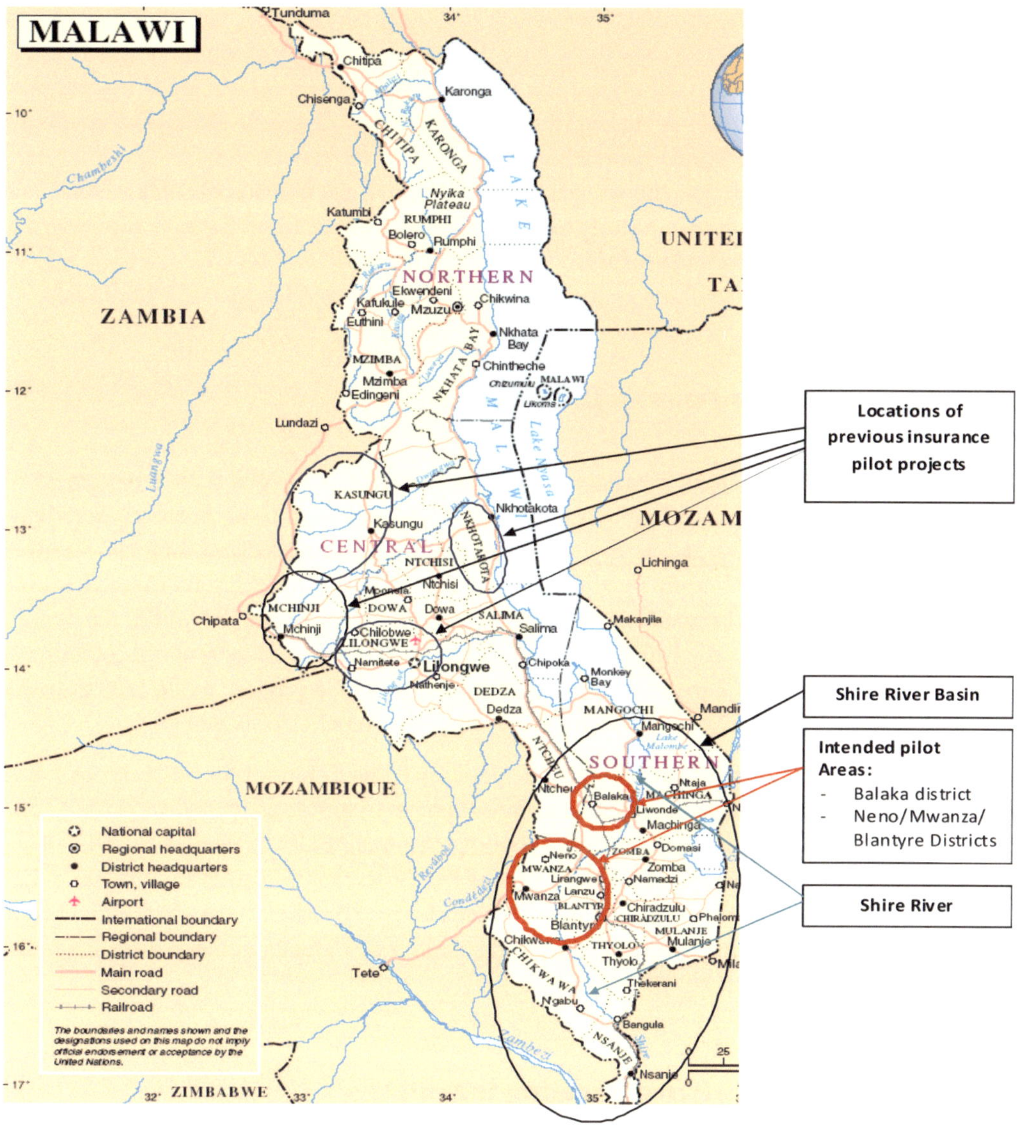

Figure 2. Selected districts of study visits.

industrial or non-agricultural sectors. Third, WII could provide realistic opportunities to commercialize and transform rural agriculture in ways that accelerate rural development in Malawi. Empirical studies suggest that WII can facilitate access to agricultural financial markets and improved farm inputs for smallholder farmers (e.g. fertilizers, high yield seed technologies, etc.). Often smallholder farmers fail to exploit the most profitable land uses and technologies due to the perceived risk and uncertainty. Farmers would rather prefer undertaking familiar and/or tested farm practices that guarantee low risk but yielding lower return and effectively limiting their potential income [3, 4]. **Figure 1** indicates the perceived rationale and justification for promoting uptake of WII as risk reduction strategy for smallholder farmers.

This paper is based on field visits, interviews and discussions with various stakeholders involved whether directly or indirectly with WII in Malawi that include government ministries, international development organizations, insurance companies, financial institutions, farmers' organizations, farmer representatives, etc. The paper discusses Malawi's experience with WII based upon on-ground observations, views and insights obtained during our interactive field visits and discussions with various organizations. Three districts of Balaka, Neno and Mwanza were selected as the main study areas—all based in Shire River Basin as shown in the map below (**Figure 2**). Other areas of study interests were districts involved in the early implementation of WII project and these included Kasungu, Lilongwe, Mchinji and Nkhotakota.

2. Weather index insurance: A false start?

As discussed earlier, WII pilot project was initiated in Malawi in 2005 with the support of the World Bank and other development partners. The project kicked off on a high note with initial involvement of 900 small scale groundnut producers who received financial support from two locally based microfinance institutions - the Opportunity International Bank of Malawi and Malawi Rural Finance Company. Within a few years after inception, the project was expanded to include maize and tobacco crops. For instance, by 2008/09 about 2600 smallholder tobacco producers were insured for a total sum of US$2.4 million, while 1700 growers (groundnuts and maize) were insured for a total sum insured of US$310,000 [6].

In these early stages, WII scheme raised high hopes as it showed signs of potential success - but only to diminish and die off a few years later. By 2014 most of the lending institutions have discontinued any form of direct crop insurance coverage to smallholder farmers. A question warranting further interrogation is 'what went wrong with a seemingly promising initiative that generated "too much hype" (borrowing Binswanger's term, [7]) and immense interest amongst development organizations, policymakers, insurance practitioners, reinsurers, academics, development specialists, etc.?'

3. Experience with WII: interrogating the pitfalls

Malawi's effort to offer unsubsidized market-driven agricultural insurance to multitude of smallholder farmers—a majority who are poor—is a laudable goal. Not only is this viewed as a climate change adaptation and mitigation strategy but also importantly a safety net against food insecurity. However Malawi has failed to implement WII beyond the pilot phase. Malawi's experience with WII therefore deserves further interrogation as this is likely to provide important insights essential for setting future research priorities and policy discourse regarding index insurance. The discussion below, for exposition purposes, is divided into three non-mutually exclusive subcomponents that include consumer-related, institution-related and infrastructure-related hurdles [8].

4. Consumer-related hurdles

4.1. Low demand for WII

A stand-alone farmer-driven demand for WII (at micro-level) has failed to take roots because of low and unsustainable demand by smallholder farmers. During the discussions with Nico Insurance companies, the general manager remarked, *"give me the demand and will market WII for you"*. The general observation is that unless insurance premiums are heavily subsidized or bundled-up with key farm inputs or loans, demand for WII is too low and unsustainable. Reasonable explanations have been offered in literature to explain the observed low demand that include: liquidity constraints, lack of trust, lack of financial and insurance literacy, high price elasticity, misconceptions of how insurance works [7, 9, 10].

Many studies suggest that a more feasible approach to offer WII is to bundle it with key farm inputs such agricultural inputs such as fertilizers, seeds, loans, etc. Key agricultural institutions (e.g. agri-banks, input suppliers, farmers' organizations, etc.) acknowledge this as an attractive approach for good reasons: can reduce default risk, can be designed to cover portfolios, can lower transaction costs and can be a strategy to reach out more clients with potential to increase business volume.

This approach has not worked well in Malawi for the reason that although farmers showed interest participating in the WII program bundled with key inputs, their insurable interest remains unknown. Rather farmers maybe participating only to exploit the 'convenience' of accessing what otherwise would have been inaccessible financial credits and/or crucial farm inputs. Minus this incentive, it's most likely that farmers will not voluntarily participate in the WII program.

The situation seem to have been worsened with the presence of a farm-input subsidy program (FISP). For the past decade the government of Malawi has rolled-out a country-wide farm input subsidy program targeting millions of smallholder farmers. The scheme is credited to have increased access to improved farm inputs (e.g. seed technology, fertilizers, chemicals, etc.) that otherwise have been too expensive and unaffordable to many smallholder farmers. During the focus group meetings, farmers indicated that they have been beneficiaries of this program since its inception.

When asked whether they would participate in WII (bundled-up with key inputs) given the presence of FISP, farmers were non-committal. Some were quick to criticize the FISP, voice strong objection in the manner in which the scheme is administered, while others professed ignorance about the WII including its benefits and how it operates. However, because beneficiaries are receiving farm inputs at low prices (and in some cases even for free), it reasonable that farmers are unlikely to participate in the WII program where they are expected to meet premium payment for a program they hardly know and neither guaranteed of a 'return' on premium payments (see discussion below). The overall impact therefore is that FISP has negatively impacted the demand for WII as a result.

4.2. Basis risk has discouraged demand

The basis risk looms large as one of the factors that has discouraged the demand for WII. Basis risk arises due to imperfect correlation between the index and the risk being trekked [11]. This can arise due to differential microclimate effects, rainfall distribution asymmetries, localized damages, poor spatial distribution of weather stations, etc. [12, 13]. Idiosyncratic risks such as pests/locusts invasion, disease outbreak, fire destruction, etc. provide typical examples. For instance, an insured farmer or a group of farmers may lose his/their entire crop due to a localized risk that affects only a small part of the village or district. Because such losses are localized and not widespread across all the insured farmers this does not trigger the index to justify indemnity payments. To an insured farmer, this is possibly misconstrued as a 'double loss'—the farmer has paid a premium, suffers a loss and still receives no compensation. As a result, WII has stirred dissatisfaction amongst farmers resulting in withdrawal of many participants. Under circumstances of this nature, Binswanger [7] argues that the insured farmer is actually worse off with insurance than without.

4.3. 'Poor return' on investment

Many smallholder farmers in Malawi hold the mistaken view that insurance premiums is some form of investment for which they are entitled to a "return on investment". Farmers do not view premium payments as risk mitigation tool per se but rather as some form of investment with expected future return. As farmers make successive payments without return (i.e. compensation/indemnity payments), insurance therefore yield a 'negative return'—and clearly a poor investment instrument [14]. This contributed somehow to farmers' disillusionment and/or dissatisfaction with WII in Malawi.

Parallel financial models in the form of 'village banks' were demonstrated as ideal financial models during focus group meetings and discussions with farmers in rural communities. 'Village banks' are popular informal "savings club" administered by a committee of trusted local members. There is no formal paperwork involved to become a member but rather all based on mutual trust, stature and self-standing within the community. Members are expected to pay monthly premiums which he/she can draw against or leave in the bank to earn 'interest'.

In one group it was illustrated that members pay MK500 per month or MK100 per week as subscriptions. These payments are mandatory and failure to meet this obligation may result in termination of membership. These village banks are of paramount importance to local households involved in schemes as members can borrow to meet urgent financial needs such as payment of school fees, funeral costs, farm input purchases, supplementary food purchases, etc. Although formal financial institutions are available (e.g. Malawi Rural Financial Services), villagers prefer dealing with village banks for good reasons - no stringent requirements to become a member, terms are negotiable and favorable, convenient proximity, no transaction costs and favorable 'interest rates'.

4.4. Premiums are too costly and unaffordable

Many smallholder farmers in Malawi are poor and cannot afford meeting premium payments. For this good reason many are not likely to purchase WII. Although subject of further research, farmers seem to prefer group-based compared to individual-based premium payments. This suggests that farmers prefer to participate in WII scheme as a group as opposed to individual households. National Smallholder Farmer's Association of Malawi (NASFAM), makes a similar observation when it noted high loan recovery rate (95–100%) payment by farmer groups as opposed to individual-owned.

There are advantages associated with the group-premium approach: administratively less cumbersome, provides peer-pressure amongst group members resulting in low default rates, provides lobbying and/or bargaining power, allow economies of size and scale which could result in cheaper interest rates or insurance premium. But the insurance contract needs to be transparent, easy to understand and with tangible benefits, so as to attract farmers participation.

Table 1 below 'confirms' the view expressed by most households in rural areas that they prefer borrowing loans from friends and family members rather use formal financial channels. Many interviewed farmers were not familiar with insurance in general and how it works. Using examples of life and automobile insurance, a small fraction agreed they have heard about insurance but lacks explicit knowledge and experience. For WII to be successful therefore, not only must there be need for massive insurance literacy campaign but also benefits must tangible, attractive and comparable to village bank schemes.

4.5. Communication hardships and cultural hurdles

WII is a new concept unheard of to many farmers in Malawi. One school of thoughts that can be designed to help farmers understand better is to translate the WII concept into vernacular language. However the translation into local languages is not straightforward and often difficult as there may be no parallel terms. This makes it difficult to market WII in easy-to-understand terms particularly to a layman with little or no prior knowledge of insurance. Such communication hardships require extraordinary effort to market WII in Malawi.

For some societies in Malawi, it is just un-cultural and against norms to discuss about future catastrophes such as funerals, accidents, burglary and disasters [16]. Doing so may be interpreted as inviting bad omen unto one's life. Breaking and demystifying such cultural beliefs

Source of loan	% of loan respondents
Friends in the village	69
Relatives and family members	56
Microfinance/banks	15
NGOs	8
Farmers' organization & clubs	4

Source: [15].

Table 1. Proportion of respondents indicating preferred sources of loan.

and stigmas will require massive educational campaigns to raise awareness and enlighten farmers on what WII is and is not about.

5. Infrastructural hurdles

5.1. Limited number of automated weather stations

Malawi has a total of 78 weather stations for which only 26 stations are automated and 36 stations contain data spanning more than 30-year period [17]. Weather risk at a particular station is actuarially characterized based on the weather data and variables recorded at that station. The resultant insurance premium is purely determined based on the availability of data recorded at a particular station. An insured farmer faces the basis risk as measurements at meteorological station do not always perfectly match or correlate with events on the farmer's plot. The 20 km radius has become 'one-size-fits-all' for WII contracts [18]. However what is observed in practice is that the 20 km radius criterion is not the most ideal for WII purposes. Osgood [17], estimated that under this criterion, 65% of the rural population falls outside the 'insurable radius' in Malawi.

5.2. Weather stations ill-suited for index insurance

The Department of Meteorological Services (DMS) bemoans the lack of an expansive network of weather stations in Malawi especially in rural areas. The DMS indicated that currently there are 23 government-owned weather stations which have been upgraded and automated. A number of the available long-established weather stations were constructed solely for aviation purposes and hence located near airport. It is therefore not surprising that these weather stations are far located from the remote farming communities located in rural areas.

For instance, as indicated to us during field trips, only two automated weather stations are found in districts of Balaka and Mwanza, while in other districts (e.g. Neno and Blantyre) there is not a single automated station. Such a sparse network of automated weather stations renders implementation of WII impossible as this not only worsens the 'basis risk' but also escalates farmers' dissatisfaction.

Countries that have successfully implemented index insurance (e.g. Mexico and India) are known to have committed substantial investment upfront towards the development of modern weather station infrastructure. Hence, substantial investment in weather station infrastructure is required upfront for WII to be scaled beyond the pilot phase in Malawi.

5.3. Short historical data

WII relies on historical and current weather data. Historical data are used as the basis for data analysis in product design and pricing. Accurate and timely weather data are key to successful implementation of index insurance products. Private players in financial markets will not engage in WII market unless they can be sure that good data are available for pricing contracts and the data are reliable and timeously provided so to settle claims quickly. The relatively high cost of private weather data services constrains the potential for scaling up insurance in many remote, rural areas with low levels of agricultural productivity, sparse populations and difficult terrain.

Unfortunately, both historical and current data are not always plentiful. The historical dataset is highly variable for different areas, particularly for daily data, which are needed for index design. In addition to poor spatial distribution, many stations do not have the long historical data, necessary for risk, hazard and exposure modeling and characterization. Some are stations are new and besides capturing precipitation, important meteorological variables (e.g. sunshine, wind speed, maximum and minimum temperatures) necessary for WII purposes are not recorded. In some instances, there data are not well archived and maintained for insurance rating and modeling purposes.

6. Institutional hurdles

6.1. Lack required expertise

Leading insurance companies acknowledge capacity constraints as the main hurdle hampering efforts to implement WII on sound footing. **Table 2** shows areas and lacking expertise as indicated by insurance companies during the field visits and discussion. Some companies have tried to utilize regional underwriting actuarial expertise from South Africa though without much success. The interviewed insurance companies expressed the view that WII is a new type of insurance that is totally different from the familiar traditional insurance (e.g. automobile insurance, life insurance, etc.). Hence for many firms it's still a "learning curve."

Banking institutions in particular, indicated WII as a costly failure in Malawi—justifying disinterest and abandonment of the program. The experience of Opportunity Bank which lost over $1.1 million after providing insurance coverage to more than 1000 small scale groundnut growers (2005), was a commonly cited example. Many institutions pointed out reasons such as lack of collateral security by smallholder farmers, high transaction costs, low insurance literacy, high default risk, non-enabling legal and regulatory framework.

6.2. Cash versus food crops conflict

Empirical studies have so far has demonstrated that the financial institutions are willing to offer insurance coverage for cash crops as opposed to food crops. This stems from purely business rationale that offering insurance coverage to cash crops protects the bank portfolios more effectively than non-cash crops. Cash crops like cotton and tobacco have a well-defined and long-established market supply-value chain in existence for many decades.

This sharply contrasts the government vested interest in providing insurance coverage to food crops especially staple maize. The government interest is understandable as it seeks to enhance resilience to climate change risks and achieve national goals and priorities especially food security, income improvement, poverty alleviation, elimination of malnutrition and diseases.

Area	Expertise lacking
Product design	• Insurance products are often designed in other countries (e.g. South Africa) and adapted to the Malawian market. Insurance products are standardized rather than tailor-made. • Insurance companies lack product design innovativeness; rather than develop and test own products, they copy products implemented and successful elsewhere • Lack of skilled staff to design products and pricing structure • Leading insurance companies has no actuarial in-house expertise
Product marketing	• No expertise within insurance sector to set up a marketing strategy for insurance products • Insurance companies not present in rural market segments, potential rural customers face difficulties to get access to insurance services
Customer service	• Insurance companies (departments) are difficult to reach • Claim settling takes a long time

Source: Authors' compilation.

Table 2. Area of lacking expertise.

7. Lessons from other developing countries

The past decade has witnessed serious attempts by developing countries in partnership with development organization (e.g. World Bank, DfID, IFAD, WFP, etc.) to develop agricultural insurance markets that currently are largely missing. Between 2007 and 2009 for instance, IFAD reports that 30 WII programmes in 19 countries were implemented, reaching more than 1.2 million beneficiaries predominantly smallholder farmers [19, 20]. Concerns for increasing uncertainty due to climate change and climate variability, have heightened interest in developing agricultural insurance markets particularly WII. This section shares in brief, lessons from other developing countries that have successfully implemented index insurance beyond the piloting phase.

7.1. Mainstream WII into nation-wide disaster risk reduction programs

For many countries that have successfully implemented WII, the program has been mainstreamed to become part and parcel of nation-wide climate change disaster risk management programs. Although WII is not a panacea, its effectiveness can be improved when combined with other risk management strategies (e.g. drought/flood early warning, seasonal forecasts, disaster-risk preparedness and etc.). It must not be viewed as a substitute for climate change adaptation but rather as an additional component of nation-wide disaster risk reduction strategies. For instance, the government of Mexico, institutionalized risk management by establishing the National Civil Protection System (SINAPROC) with a nation-wide mandate to mitigate the loss of lives and interruption of essential services caused by disaster [21]. To further strengthen its risk management strategy, the Mexican government established a National Fund for Natural

Disasters (FONDEN) in 1996 specifically intended to cater for the reconstruction of public infrastructure as well as offering compensation to low income smallholder producers.

7.2. WII must create value to farmers

WII must create value to farmers in tangible a manner as possible. Farmer will value WII as a credible risk management strategy if it can effectively lower vulnerability to climate change, improve farm income, alleviate food insecurity and poverty. According to Churchill [22], the concept of value includes: scope of coverage - the extent to which the farmers' actual risk is covered; affordability - the extent to which the products are financially accessible to the farmers; additional benefits - the extent to which the insurance gives access to any additional services that might make the farmers "better off." In other words, WII products are deemed valuable if they are accessible, appropriate and affordable enough to allow participation of low income farmers (Matul et al. 2013). **Table 3** provides the PACE model considered vital to enhance the value of index insurance in ways that positively impacts the demand for WII by farmers.

7.3. Build efficient product delivery channel premised on trust

Patt et al. [23] (cited in Matul et al., [24] suggests, as a best practice, that practitioners need to implement a three-pronged approach to building trust: 1) build trust in the product, 2) build trust in the insurer and other institutions involved in the delivery of the product, especially through use of trusted agents and messengers, and 3) leverage trust that already exists in communities. For instance it could be easy to build trust on products delivered with the involvement of extension officers who consistently work with farmers. What often makes it difficult is that most insurers normally do not have offices or business in rural areas. This means that product distribution is best handled through a party with extensive links with farmers or farmer groups (e.g. extension officers, banks, agro-processors, cooperatives or MFIs). For instance in Kenya, WII products are being promoted as bundled products with loan inputs and marketed via agro-input dealers.

Component	Description
Product	describes appropriateness by reviewing coverage, benefit level, eligibility criteria and availability of value-added services
Access	focuses on accessibility and simplicity by investigating choice, enrolment, information, education, premium payment method and proximity
Cost	measures both affordability and value for money, while looking at additional costs to keep down overall costs of delivery
Experience	assesses responsiveness and simplicity by looking at claims procedures and processing time, policy administration, product tangibility and customer care

Source: Matul et al. [24].

Table 3. Value of WII based on PACE model.

7.4. Provide smart subsidies necessary especially during early stages of WII

Smart subsidies may be are necessary especially during early stages of WII. Both developed and developing countries that have succeeded in setting up strong crop insurance schemes (e.g. India, Mexico, China, etc.), show that this success has been due, in large part, to public support. These countries have provided support mostly through direct provision of premium subsidies and reinsurance.

Africa has seen very low levels of government subsidies and, consequently, very low levels of market development, particularly for small-scale farmers. Statistics on agricultural insurance show that only 220,000 households were covered by agricultural microinsurance in Africa by 2011, representing US$6.61 million in premiums [25]. Malawi (likewise other African countries) need to learn especially from India and Mexico that developmental path of agricultural insurance market starts with heavy government support and only for the private sector to join after the WII programs have fully matured and passed the "market test." Hazell et al. [26] argue there may be good reasons for subsidizing development-oriented WII for the very poor, especially if it displaces more costly types of safety nets.

However, subsidies are always controversial. They are costly and unsustainable and if not carefully considered, subsidies can form a huge fiscal drain on national budgets with far reaching distortions on insurance markets. As Sandmark et al. [25] emphasize, subsidy support has to be weighed against market efficiency and fiscal sustainability. Experience indicates that once established, subsidies may be difficult to eliminate. **Table 4** indicates the pros and cons of providing subsidies to WII programs.

	Description
Pros	• Can make insurance products affordable to the very poor • Can provide incentives for WII uptake • Can jump-start the market • Can be designed so as decrease as uptake increases • Subsidized services can decrease the need for other types of social safety nets
Cons	• Often provided indiscriminately for all clients, when some clients are able to pay full (or higher) premiums • Can promote dependency culture and provide disincentives to purchase non-subsidized products • Can depress the market • Can crowd out private insurers that do not receive subsidies

Source: [7].

Table 4. The pros and cons of providing subsidies to WII.

7.5. Invest in national weather services and infrastructure

One reason behind India's successful WII story is its massive investment in improving weather station infrastructure. According to Hazell et al. [26], historical data are only available from approximately 550 India Meteorological weather stations. These are far too few to adequately cover the 150 million hectares of arable land and rarely located in rural areas. It has been estimated that so far between10,000–15,000 additional weather stations have been installed so as to offer meaningful insurance services to farmers. Of these, at least 5000 are automatic, each at a distance of approximately 10–20 km from insured land [26].

Beyond the investment in weather stations, there is a need to collect, maintain and archive data - making them available on real time in relation to insured events. World Meteorological Organization (WMO), private players, donors and governments—working closely with national meteorological services—can play a key role in improving and expanding the network of weather stations and the quality of data produced and available. Further, these actors can explore opportunities related to satellite-based indices that use remote sensing tools. Given the increasing unpredictability of weather patterns due to the changing climate, the benefits of investment in weather infrastructure will extend beyond the development of index insurance products [26].

7.6. Intensify insurance literacy and financial education campaigns

Many smallholder farmers have limited experience with formal insurance. Educating farmers on financial risk management is necessary to enhance insurance literacy [25]. Farmers require educational drills on agricultural insurance (e.g. WII) as a risk reduction, diversification and protection tool—all which will help farmers develop better expectations of WII.

Colombia's approach to financial education on insurance provides a good example. In 2009 the government made financial education on insurance mandatory for all microfinance institutions (MFIs). As a result, MFIs embarked on massive program to educate their clients through workshops, radio and television programs, leaflet distribution and internet campaigns. **Figure 3** shows the steps involved in financial literacy campaign in Colombia starting with "raising awareness" of risk to the final goal being to "change people's behaviour" towards risk, and allow they to make informed decision and ultimately reduce the impact of the risks.

7.7. Intensify the use of mobile phones and/or technology

The use of mobile phones adds an exciting dimension to WII as this has the potential to lower administrative costs. Mexico, Brazil and Kenya provide good examples of how to use technology to improve distribution channels and increase outreach of insurance. Kenya, for instance, mobile technology called M-PESA is widely used in performing all paperless transactions including registration, premium payment and indemnity pay-out and has brought great benefits to the insurance industry (Bankable Frontier [27]). The M-PESA system has improved insurance penetration especially in rural areas as it reaches even those clients in remote parts.

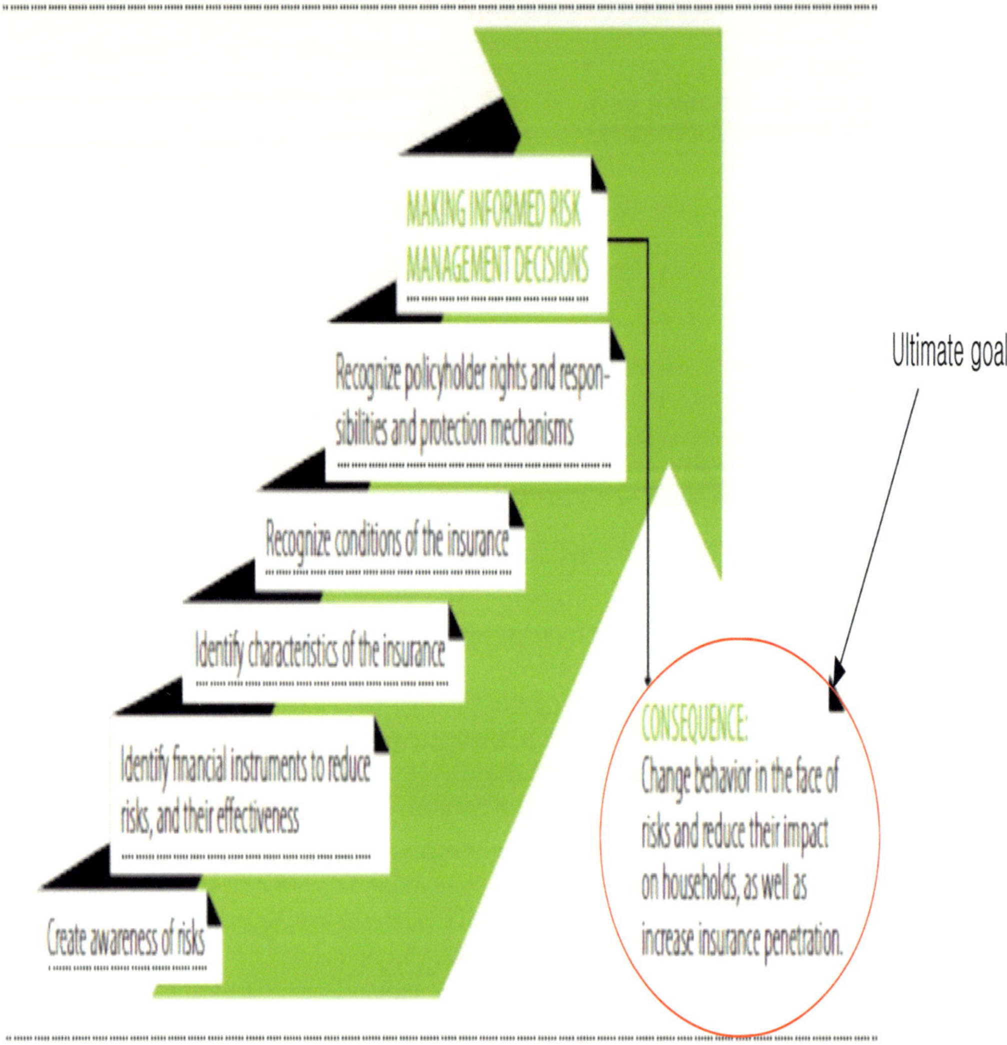

Figure 3. Continuum of financial education on insurance in Colombia. Source: Vercla and Carlos, 2013.

Mobile technology is an important tool for WII as reduces the cost and increase the efficiency of the index insurance programs. One of the ways this is seen in practice is in distribution, where more efficient processes reduce the costs of distribution. Likewise, greater outreach may achieve a product scale that, in turn, drives down costs. But for technology to function effectively, it must meet the needs of the user and be adaptable to the environment in which it works.

8. Conclusion

WII holds promise as a one of the disaster risk reduction strategies in Malawi important for mitigating the impact of extreme weather events - droughts and floods in particular. Despite its potential, weather index insurance has failed to develop roots and grow beyond the pilot phase.

Based on field visits and discussions with key organizations involved in implementation of WII, this study has discussed some of the hardships that hampered successful development of WII in Malawi. The study also draw on a few illustrations (e.g. Mexico, India, Colombia and Kenya) to demonstrate successful experience and immense potential of WII as a risk protection strategy. Drawing from this review, the study identified some key lessons and best practice that could help the government of Malawi to re-visit, re-think and revise WII program as disaster risk reduction strategy against extreme weather events imposed by climate change and climate variability.

Acknowledgements

The author is grateful to individual farmers, farm organizations, government officials (with various ministries and/or departments), insurance firms, financial institutions and other stakeholders that we interacted with and interviewed during the course of the study in Malawi. I would like to express gratitude to Mr. Wim van den Bos who participated as one of the research team. Finally, I acknowledge financial and material support provided by UNDP during the course of the study. The views, gaps, omissions and errors in this report belong to the author and have nothing to do with UNDP as an organization.

Author details

Ephias M. Makaudze

Address all correspondence to: emakaudze@gmail.com

Malone University, Canton, Ohio, USA

References

[1] Pauw K, Thurlow J, Seventer D. Droughts and Floods in Malawi—Assessing the Economicwide Effects. IFPRI Discussion paper 00962; 2010

[2] World Bank. Malawi 2015 Floods post Disaster Needs Assessment Report. Global Facility for Disaster Reduction and Recovery; 2015

[3] Makaudze E, Miranda M. Catastrophic drought insurance based on remotely sensed normalized difference vegetative condition index for smallholder farmers in Zimbabwe. Agrekon. 2010;**49**(4):418-432

[4] Skees JR, Collier B. The Potential of Weather Index Insurance for Spurring a Green Revolution in Africa. Lexington: GlobalAgRisk Inc.; 2008

[5] Cole S, Bastian G, Vyas S, Wendel C, Stein D. The Effectiveness of Index Based Micro-Insurance in Helping Smallholders Manage Weather-Related Risks. London: EPPI-Centre, Social Science Research Unit, Institute of Education, University of London; 2012

[6] Mapfumo S. Weather Index Based Insurance in Malawi, paper presented in Uganda; 2009. Available from: www.siteresources.worldbank

[7] Binswanger-Mkhize H. Is there too much hype about index-based weather index insurance. Journal of Development Studies. 2012;**48**:187-200

[8] Makaudze E, editor. Weather Index Insurance for Smallholder Farmers in Africa: Lessons learnt and goals for the future. In: Conference Proceedings, Stellenbosch: SunMedia; 2012. ISBN: 978-1-919985-45-9

[9] Cole SA, Gine X, Tobacman J, Topalova P, Townsend R, Vickery IJ. Barriers to household risk management: Evidence from India, Working Paper No. 09-116. Harvard Business School Finance; 2009. FRB of New York Staff Report No. 373

[10] Clarke DJ, Mahul O, Rao KN, Verma N. Weather Based Crop Insurance in India, Policy Research Working Paper. Washington, DC: World Bank; 2012. pp. 31

[11] Clarke A. Insurance Design for Developing Countries. [Ph.D. diss. thesis]. Oxford University; 2011

[12] Miranda MJ, Farrin K. Index insurance for developing countries. Applied Economic Perspectives and Policy. 2012:391-427

[13] Lebrois L, Quirion P. Agricultural insurances based on meteorological indices: realizations, methods and research challenges. Meteorological Applications. 2013;**20**:1-9

[14] Slovic P, Fischhoff B, Lichtenstein S, Corrigan B, Combs B. Preference for insuring against probable small losses: Insurance implications. The Journal of Risk and Insurance. 1977;**44**:237-258

[15] UNDP. Private Public Sector Partnership on Capacity Building for the Sustainable Land Management in the Shire River Basin. Lilongwe; 2010

[16] Livata G. Using Index-Based Weather Insurance – The Malawi Experience, Expert Meeting on Risk Management for Financing to the Agriculture Value Chain in Africa, Opportunity International Bank of Malawi. 1-3 April 2009

[17] Osgood D, Suarez P, Hansen J, Carriquiry M, Mishra A. Integrating Seasonal Forecasts and Insurance for Adaptation among Subsistence Farmers: The Case of Malawi. World Bank Policy Research Working Paper 4651. 2008

[18] Greatrex H, Grimes DIF. Capturing the spatial variability of rainfall in weather-based index insurance. In: The challenges of index-based insurance for food security in developing countries. Proceedings of a technical workshop organised by the EC Joint Research Centre (JRC) and the International Research Institute for Climate and Society (IRI, Earth Institute, Columbia University), JRC Ispra, Italy, 2 and 3 May 2012. R Gommes and F. Kayatikire eds.

[19] Ceballos F, Roble M. Weather risks and insurance opportunities for the rural poor. Building resilience for food and nutrition security. Addis Ababa: 2020 Conference; 2014

http://dx.doi.org/10.5772/intechopen.77106

[20] International Fund for Agricultural Development (IFAD). Weather Index-based Insurance in Agricultural Development: A Technical Guide. [Online]; 2011. Available from: www.ifad.org/ruralfinance/pub/wii_tech_guide.pdf

[21] Ibarra H, Syroka J. Case studies for agricultural weather risk management. In: Castiglione E, editor. Risk Management in Agriculture for Natural Hazards, Rome: ISMEA. 2006

[22] Churchill C, Matul M, editors. Protecting the poor – A Microinsurance compendium'. Vol II; ILO, Switzerland. 2012, ISBN: 978-92-2-125744-8

[23] Patt GA, Peterson N, Carter M, Velez M, Hess U, Suarez P. Making index insurance attractive to farmers. Mitigation and Adaptation Strategies for Global Change. 2009;14(8):737-753

[24] Matul M, Jaleran CT, Kelly E. Improving Client Value from Microinsurance: Insights from India, Kenya and the Philippines. Geneva: ILO; 2011

[25] Sandmark T, Debar JC, Tatin-Jaleran C. The Emergence and Development of Agriculture Microinsurnce. A Discussion paper. Lexembourg: Microinsurance Network. [Online]; 2013. Available from: http://www.fondation-farm.org/zoe/doc/micro_network-brochure_agriculture-def-low.pdf

[26] Hazell P, Anderson J, Balzer N, Hastrup Clemmensen A, Hess U, Rispoli F. The Potential for Scale and Sustainability in Weather Index Insurance for Agriculture and Rural Livelihoods. Rome: International Fund for Agricultural Development and World Food Programme; 2010

[27] Bankable Frontier Associates. Review of FSD's index based weather insurance initiatives. Nairobi: Financial Sector Deepening; 2013